Exam Practice
A LEVEL

A Level
Exam Practice

Covers AS and A2

Mathematics

Author

Michael Jennings

Contents

AS and A2 exams

Different types of questions

Questions on mathematics papers have varying degrees of structure to them. Those at the beginning of a paper tend to be short and sharp and worth only a few marks and thus there is little opportunity to break the question down into smaller parts. However, as you progress through the paper, the questions become longer and more challenging and will often have several parts to them. These parts could be totally independent or there may be a common theme running through a question where the examiner is attempting to lead you through the early parts in order to give you a hint on a method that might be used to do a later, more difficult part – the wording "hence, or otherwise" is an indication of this type of question.

A question can also be broken down into smaller parts in order for a candidate who gets stuck on an early part to, nonetheless, be able to go on and score marks on later parts. To this end, the answer may be given in a particular part and the candidates be asked to show it. Thus if a candidate is unable to derive the correct result, he or she can use the printed answer to hopefully progress further through the question.

Sometimes candidates may be required, as the first part of a question, to produce a proof or derivation of a standard result or formula which is then used in a subsequent part or parts to solve a particular problem. Explanations or definitions of particular terms may also be asked for, particularly on statistics papers.

You may also be asked to comment on or interpret a result (in statistics or decision mathematics), explain a modelling assumption and where it is used (in mechanics), identify a flaw in an argument (in pure mathematics) or how a particular model could be refined to make it more realistic.

What examiners look for

- clear and concise methods, although any valid method, no matter how long, will be given full credit.
- appropriate and accurate use of notation and symbolism.
- large and clearly labelled diagrams and graphs where appropriate.
- appropriate and accurate use of technology (e.g. a calculator).
- the ability to interpret and comment on results obtained.

What makes an A, C and E grade candidate?

- **A grade candidates** have a broad knowledge of mathematics and can apply that knowledge in a wide variety of situations, including unfamiliar scenarios, accurately and efficiently. They are strong on all of the units. The minimum mark for a grade A is 80% on the Uniform Mark Scale.
- **C grade candidates** have a fair knowledge of mathematics but find it less easy to apply their knowledge in unfamiliar situations. Their work is less accurate and they have weaknesses on some of the units. The minimum mark for a grade C is 60% on the Uniform Mark Scale.
- **E grade candidates** have a poor knowledge of mathematics and are unable to apply it in unfamiliar situations. Their work has many errors and they are unable to recall key facts and techniques. The minimum mark for a grade E is 40% on the Uniform Mark Scale.

Successful revision

Revision skills

- By far the best way to revise for mathematics is by doing mathematics i.e. by solving problems. Of course you have to learn the theory but unless you can apply the theory to actually tackle problems, the knowledge is of little use. It is therefore essential, if your preparation is to be effective, that you encounter as many different situations and scenarios as possible, by doing as many practice questions as possible, so that you can learn how to recognise which techniques will be appropriate to solve a particular type of problem and which will not.

- When revising the theory, try to summarise, as concisely as possible, the key points and how they relate to other parts of the syllabus. Writing out your own concise notes (the briefer the better) for each syllabus topic can be a good way of learning material.

Practice questions

To use this book effectively

- Examine the grade A and grade C sample answers and make sure that you understand where the errors have been made and how to correct them.

- Try the exam practice questions – don't be tempted to look at the answers too quickly if you get stuck; you will learn a great deal more from a question if you struggle with it and eventually sort it out or at least make some progress, by yourself, using worked examples in your notes or in a textbook to guide you.

- When you feel confident and ready, try the mock exam papers.

Common errors

Many errors occur due to careless work with signs, particularly when removing brackets, and errors in basic algebra and trigonometry.

1. **Many of the most common errors occur as a result of students treating all functions, f, as being linear**

   i.e. $f(a + b) = f(a) + f(b)$,   for all a and b.

   e.g.

   $(a + b)^2 = a^2 + b^2$, or similar

   $\sqrt{(a+b)} = \sqrt{a} + \sqrt{b}$

   $\dfrac{1}{a + b} = \dfrac{1}{a} + \dfrac{1}{b}$

   $\sin (A + B) = \sin A + \sin B$

   $\ln (A + B) = \ln A + \ln B$

   $e^{x+y} = e^x + e^y$

   Of course none of the above are true, **in general**; (some of them *may* be true in certain special cases). See if you can, where possible, correct them.

2. **Confusion with notation**

   e.g.

   f^{-1}, the **inverse** of f, is often confused with f', the **derivative** of f

   fg means "do g first then f", not the other way round.

How to boost your grade

- Ensure that you do exactly as the question says, e.g. if you are told to use a particular method then you will receive no credit whatsoever for using a different method, even if you get the question right.

- Ensure that you give answers to the correct degree of accuracy when requested to do so – you will definitely lose marks if you don't.

- Show your working – a very high proportion of the available marks at A level are Method Marks.

- You can answer the questions in any order that you like – you should attempt a few of the shorter questions at the beginning of the paper to boost your confidence, making sure that you leave yourself plenty of time for the last two or three questions, for which there are a very high number of marks.

- For the shorter questions, make life easier for the examiner, by ruling off at the end of a question and either leave a space before you start the next question, or if you are near the bottom of the page, start on a fresh piece of paper. Always start the longer questions on a fresh page. This will help to avoid copying and transcription errors which are made when turning over a page.

- If you attempt a question using two different methods, then do not cross either of them out but instead leave both – the examiner will mark both and award you the better mark.

- Dimensional analysis, particularly in algebraic Mechanics questions, will often help to spot silly mistakes. For example, if you are asked to find the loss in kinetic energy in a particular problem and you obtain an answer of $5mu$, you should realise that you have made an error as this expression has momentum (or impulse) units.

- Don't work *too* quickly – try to check each line of working before moving on to the next – but on the other hand don't waste time e.g. by underlining everything; use your time sensibly – try to match the time that you use to the marks available for that question – if you get stuck on a question, particularly a short one, don't panic! Leave a space and go back to it later, if you have time.

- Familiarise yourself with the Formulae Booklet before you do the exam – make sure that you know what is in there and where it is situated.

- Make sure that you have a calculator with you, that it is a permitted one (you are not allowed graphics calculators for certain pure units), and that it works!

- Put all your past papers together and look through them so that you are familiar with the type of questions that are asked and look through your copy of the syllabus to make sure that you have revised all the topics.

Glossary of terms used in examination questions

Prove – Show that a result is true, using a reasoned argument which starts from accepted basic results (the question will sometimes clarify what you can assume)

Write down, state – no justification is needed for your answer.

Calculate, find, determine, show, solve – show sufficient working to make your method clear. (N.B. Answers without working will gain no credit)

Deduce, hence – use the given result or previous part to establish the result.

Sketch – graph paper not needed; show the general shape of a graph, where it crosses the axes (if it does), any asymptotes and any points of particular significance.

Draw – plot accurately on graph paper using a suitable scale.

Find the exact value – leave your answer as a fraction or in surds, or in terms of logarithms, exponentials or π; note that using a calculator is likely to introduce decimal approximations, resulting in a loss of marks.

Questions with model answers

For help see Revise AS
Study Guide pages 30–32

C grade candidate – mark scored 7/13

N.B. Only a basic scientific calculator may be used for P1 questions

(1) The straight line l passes through the point (–2, 1) and is perpendicular to the line m with equation $2y - x + 11 = 0$.

(a) Find an equation for l. **[6]**

Equation of m is $y = \frac{1}{2}x - \frac{11}{2}$ → gradient = $\frac{1}{2}$

→ gradient of l is 2

Equation of l is $y - 1 = 2(x - -2)$

→ $y = 2x + 5$.

(b) Find the coordinates of the point where the lines l and m intersect. **[4]**

$y = \frac{1}{2}x - \frac{11}{2}$ and $y = 2x + 5$

i.e. $\frac{1}{2}x - \frac{11}{2} = 2x + 5$

→ $\frac{3x}{2} = \frac{21}{2}$

→ $x = 7$ → $y = 19$.

(c) Verify that the point $A(5, -3)$ lies on the line m. **[1]**

$(2x - 3) - 5 + 11 = 0$.

(d) Deduce the perpendicular distance of A from the line l. **[2]**

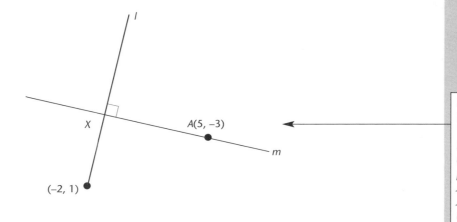

Examiner's Commentary

Correct; has put in form $y = mx + c$ and read off the value of m.
Has correctly inverted but forgotten to change the sign to obtain the gradient of the normal.
Correct use of $y - y_1 = m(x - x_1)$ but wrong value of m.
Incorrect answer, **4/6 scored.**

Correct substitution method.

Another sign error!

Incorrect, **2/4 scored.**

An easy mark, **1 scored.**

No answer given, **0 scored.**
The idea is that since the lines intersect at right angles at X, say, the perpendicular distance of A from l is the length of AX, where X should be (–7, –9) (see diagram).

Questions with model answers

? For help see Revise AS Study Guide pages 21 and 28

A grade candidate – mark scored 9/11

(2) **(a)** Find the values of the constants a, b and c such that
$a(x + b)^2 + c = 4x^2 - 24x + 27$, for all x. **[4]**

$$a(x^2 + 2bx + b^2) + c = 4x^2 - 24x + 27$$

$$a = 4; \ 8b = -24, \ b = -3;$$

$$4b^2 + c = 27, \ 36 + c = 27, \ c = -9.$$

Examiner's Commentary

Brackets correctly expanded.

Coefficients of x^2 and x equated and the constants.

(b) Hence, or otherwise, find the set of values of x
for which $4x^2 - 24x + 27 \geqslant 0$. **[3]**

$$4(x - 3)^2 - 9 \geqslant 0,$$

$$x - 3 \geqslant \frac{3}{2} \ or -\frac{3}{2}$$

$$x \geqslant \frac{9}{2} \ or \ \frac{3}{2}.$$

This is an easy mistake to make; particular care needs to be taken when square rooting inequalities.
Should be:
$x \geqslant \frac{9}{2}$ *or* $x \leqslant \frac{3}{2}$,
2/3 scored.

(c) Sketch the graph of $y = 4x^2 - 24x + 27$, showing where it cuts the axes and the coordinates of any turning points. **[4]**

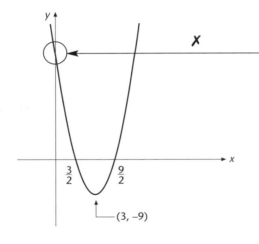

Omitted this value.
3/4 scored.

Exam practice questions

A *Answers on pp. 15–21*

1.1 Algebra

(1) The roots of a quadratic equation $x^2 + bx + c = 0$ are $2+\sqrt{5}$ and $2-\sqrt{5}$.
Find the values of b and c. **[5]**

(2) The polynomial $f(x) = 2x^3 + px^2 - x + q$ is exactly divisible by $(x - 1)$
and $(x + 2)$.

    (a) Find the values of p and q. **[5]**

    (b) Hence factorise $f(x)$. **[2]**

(3) Solve the simultaneous equations $\quad 2x - y = -1$
$\qquad\qquad\qquad\qquad\qquad\qquad\quad xy + y = 3.$ **[5]**

1.2 Coordinate geometry

(1) The point A has coordinates (−4, 2) and the point B has coordinates (−2, 4).

    (a) Find an equation of the straight line l which passes through the origin
          O and the mid-point of AB. **[4]**

    (b) Find an equation of the straight line m which passes through B and
          the mid-point of OA. **[2]**

    (c) Find the point of intersection of the lines l and m. **[2]**

(2) The straight line l passes through the points P and Q with coordinates
(2, 2) and (6, 0) respectively.

    (a) Find an equation of l. **[4]**

The straight line m passes through the point R with coordinates (−9, 0) and has
gradient $\frac{1}{4}$.

    (b) Find an equation of m. **[2]**

The lines l and m intersect at the point S.

    (c) Calculate, to 2 decimal places, the length of PS. **[3]**

    (d) Calculate the area of $\triangle SRQ$. **[2]**

Exam practice questions

(3)

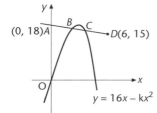

The graph shows part of the curve with equation $y = 16x - kx^2$, where k is a constant. The points A and D have coordinates $(0, 18)$ and $(6, 15)$ respectively.

(a) Calculate, giving your answer to 3 significant figures, the length of AD. **[2]**

The line l passes through the points A and D and intersects the curve at the points B and C, as shown.

(b) Obtain an equation of l in the form $y = mx + c$, where m and c are constants. **[4]**

Given also that C has coordinates $(4, 16)$,

(c) show that $k = 3$, **[2]**

(d) calculate the coordinates of B. **[4]**

1.3 Sequences and series

(1) An arithmetic series has 3rd term -20 and 11th term 20.

(a) Find the first term and the common difference. **[5]**

The sum of the first k terms of the series is zero.

(b) Find the non-zero value of k. **[4]**

(2) Mark is given an interest-free loan to buy a second-hand car. He repays the loan in monthly instalments. He repays £20 the first month, £22 the second month and the repayments continue to rise by £2 per month until the loan is repaid. Given that the final monthly repayment is £114,

(a) show that the number of months that it will take Mark to repay the loan is 48, **[3]**

(b) find, in pounds, the price of the car. **[4]**

(3) A sequence of numbers $u_1, u_2, \ldots, u_n, \ldots$ is given by the formula $u_n = 3\left(\frac{2}{3}\right)^n - 1$, where n is a positive integer.

(a) Find the values of $u_1, u_2,$ and u_3. **[2]**

(b) Find $\sum\limits_{n=1}^{15} 3\left(\frac{2}{3}\right)^n$, and hence show that $\sum\limits_{n=1}^{15} u_n = -9.014$ to 4 significant figures.

[5]

(c) Prove that $3u_{n+1} = 2u_n - 1$. [4]

(4) The third term of a geometric series is 15 and the common ratio of the series is 2.

(a) Find the sixth term of the series. [2]

(b) Find the sum of the first ten terms of the series. [3]

The second and fifth terms of the series form the first two terms of an arithmetic series.

(c) Find the ninth term of the arithmetic series. [5]

(d) Find the sum of the first thirteen terms of the arithmetic series. [3]

(5) The first, second and third terms of a geometric series are $(x + 4)$, $(x + 1)$ and x, respectively.

(a) Find the value of x. [5]

(b) Find the common ratio of the series. [2]

(c) Find the sum to infinity of the series. [2]

(6) A 'Yearly Plan' is a National Savings scheme requiring 12 monthly payments of a fixed amount of money on the same date each month. All savings earn interest at a rate of $p\%$ per complete calendar month.

Alex decides to invest £30 per month in this scheme and makes no withdrawals during the year.

(a) Show that, after 12 complete calendar months, his first payment has increased in value to £$30r^{12}$, where $r = 1 + \frac{p}{100}$. [4]

(b) Show that the total value, after 12 complete calendar months, of all 12 payments is $\dfrac{£30r(r^{12} - 1)}{(r - 1)}$. [4]

(c) Hence calculate the total interest received during the 12 months when the monthly rate of interest is $\frac{1}{2}$ per cent. [4]

(7) Find $\sum\limits_{r=1}^{10} (r + 3^r)$. [6]

Exam practice questions

1.4 Trigonometry

(1) Solve, for $0° \leqslant x < 360°$, the equation $3\cos^2 x - 2\sin x = 2$. **[7]**

(2) The curve with equation $y = 3 + k\sin x$ passes through the point with coordinates $(\frac{\pi}{2}, -1)$. Find

    (a) the value of k **[2]**

    (b) the greatest value of y. **[2]**

(3) Given that $0 \leqslant x \leqslant \pi$, find the values of x for which

    (a) $\cos 3x = 0.5$, **[4]**

    (b) $\tan(x + \frac{\pi}{2}) = -1$. **[3]**

(4) **(a)** Given that $\sin 15° = \dfrac{(\sqrt{6} - \sqrt{2})}{4}$, find in the form $\sqrt{m + n\sqrt{3}}$,

    where m and n are rational, the value of

    (i) $\cos 15°$ **(ii)** $\sin 105°$. **[6]**

    (b) Find, in radians to two decimal places, the values of x in the interval $0 \leqslant x \leqslant 2\pi$, for which
$$3\cos^2 x + \cos x - 2 = 0.$$ **[6]**

(5) In the diagram, O is the centre of the circle and AB is a chord. The radius of the circle is 8 cm and $A\hat{O}B = 120°$.

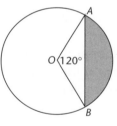

Calculate:

    (a) the perimeter of the shaded region **[6]**

    (b) the area of the shaded region. **[5]**

1.5 Differentiation

(1) Find the coordinates of the turning points on the curve whose equation is
$$y = x^3 - 9x^2 + 24x.$$ **[11]**

(2) A rectangular tank is made of thin sheet metal. The tank has a horizontal square base, of side x cm, and no top. When full the tank holds 500 litres.

    (a) Show that the area, A cm², of sheet metal needed to make this tank is given by
$$A = x^2 + \frac{2\,000\,000}{x}, \quad x \neq 0.$$ **[6]**

(b)   Find the value of x which makes A a minimum and find this minimum value of A.   **[6]**

(c)   Prove that this value of A is a minimum.   **[4]**

(3)   A curve has equation $y = 2x - x^2$

(a)   Find an equation of the normal to the curve at the origin O.   **[5]**

(b)   Find the coordinates of the point where the normal cuts the curve again.   **[5]**

(4)   The function f is given by
$$f(x) = x + \frac{1}{4x}, \ x \neq 0.$$

(a)   Find the values of x for which $f(x) = -\frac{5}{4}$.   **[4]**

(b)   Find the range of values of x for which f is an increasing function of x.   **[7]**

1.6 Integration

(1)

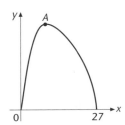

The diagram shows a sketch of the curve whose equation is
$$y = 27x^{0.5} - x^{1.5} \text{ for } 0 \leqslant x \leqslant 27.$$

(a)   Show that $dy/dx = 1.5x^{-0.5}(9 - x)$.   **[2]**

The curve has a turning point at the point A.

(b)   Find the coordinates of A.   **[3]**

(c)   Find the area of the finite region bounded by the curve and the x-axis.   **[5]**

Exam practice questions

(2)

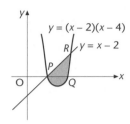

The diagram shows the line with equation $y = x - 2$ meeting the curve with equation $y = (x - 2)(x - 4)$ at the points P and R.

(a) Write down the coordinates of P and Q. [2]

(b) Find the coordinates of the point R. [4]

(c) Find the area of the shaded region bounded by the line, the curve and the x-axis. [6]

(3)

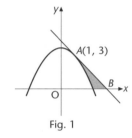

Fig. 1

The tangent to the curve with equation $y = 4 - x^2$ at $A(1, 3)$ meets the x-axis at the point B, as shown in Fig. 1.

(a) Find the x-coordinate of B. [7]

(b) Find the area of the shaded region. [9]

(4) (a) Find the general solution of the differential equation
$$dy/dx = 2(\sqrt{x - 1})^2.$$

Given that when $x = 9$, $y = 2$, [5]

(b) find y in terms of x. [3]

Answers

1.1 Algebra

(1)   Equation must factorise as

$$\{x - (2 + \sqrt{5})\}\,\{x - (2 - \sqrt{5})\} = 0$$
$$x\,\{x - (2 - \sqrt{5})\} - (2 + \sqrt{5})\,\{x - (2 - \sqrt{5})\} = 0 \quad \longleftarrow \quad \textit{Multiply out the brackets.}$$
$$x^2 - x\,(2 - \sqrt{5} + 2 + \sqrt{5}) + (2 - \sqrt{5})\,(2 + \sqrt{5}) = 0$$
$$x^2 - 4x - 1 = 0 \quad \longleftarrow \quad \textit{Collect terms.}$$
$$\text{So,} \quad b = -4 \text{ and } c = -1. \quad \longleftarrow \quad \textit{Equating coefficients.}$$

(2) (a)
$$f(1) = 2 + p - 1 + q = 0 \rightarrow p + q = -1 \quad \longleftarrow \quad \textit{Using the Factor Theorem.}$$
$$f(-2) = -16 + 4p + 2 + q = 0 \rightarrow 4p + q = 14 \quad \longleftarrow \quad \textit{Using the Factor Theorem.}$$
$$\text{subtracting gives } 3p = 15,\ p = 5 \text{ and } q = -6.$$

(b)
$$f(x) = (x - 1)(x + 2)(2x + 3). \quad \longleftarrow \quad \textit{By inspection of the constant terms.}$$

(3)
$$y = 2x + 1 \quad \longleftarrow \quad \textit{It is much easier to substitute for y.}$$
$$x\,(2x + 1) + 2x + 1 = 3$$
$$2x^2 + 3x - 2 = 0 \quad \longleftarrow \quad \textit{Multiply out and collect terms.}$$
$$(x + 2)\,(2x - 1) = 0 \quad \longleftarrow \quad \textit{Factorise or use formula.}$$
$$x = -2 \text{ or } \tfrac{1}{2}$$
$$y = -3 \text{ or } 2.$$

1.2 Coordinate geometry

(1) (a)   Mid-point of AB is $\left(\dfrac{-4 + -2}{2}, \dfrac{2 + 4}{2}\right)$ i.e. $(-3, 3)$;

therefore the gradient of $l = \dfrac{3}{-3} = -1$.

Hence, an equation of l is $y = -x$. $\quad \longleftarrow$

Notice that 'an' is used since it isn't the only one, e.g. $2y = -2x$.

(b)   Mid-point of OA is $(-2, 1)$; hence the line m has equation $x = -2$.

(c)   Solving $y = -x$ and $x = -2$ simultaneously gives the point $(-2, 2)$.

(2) (a)   Gradient of $PQ = \dfrac{(2 - 0)}{(2 - 6)} = -\dfrac{1}{2}$;

equation of l is $y - 0 = -\dfrac{1}{2}(x - 6)$ i.e. $2y + x - 6 = 0$.

(b)   Equation of m is $y - 0 = \dfrac{1}{4}(x - -9)$ i.e. $4y - x - 9 = 0$.

(c)   Find S by solving simultaneously: $2y + x - 6 = 0$
$$4y - x - 9 = 0$$
$$\text{adding gives} \quad 6y - 15 = 0$$
$$y = \tfrac{5}{2} \rightarrow x = 1 \text{ i.e. } S \text{ is the point } \left(1, \tfrac{5}{2}\right)$$

Hence $PS = \sqrt{(2 - 1)^2 + \left(2 - \tfrac{5}{2}\right)^2} = \sqrt{\tfrac{5}{4}} = 1.12$ (2 d.p.)

(d)   Draw a simple diagram: Area of $SRQ = \tfrac{1}{2}.15.\tfrac{5}{2} = \tfrac{75}{4}$ sq. units.

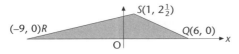

$S(1, 2\tfrac{1}{2})$

$(-9, 0)R$ $\qquad$ $Q(6, 0)$

O $\qquad x$

Answers

(3) (a) $AD = \sqrt{(0-6)^2 + (18-15)^2} = \sqrt{45} = 6.71$ (3 s.f.)

(b) Gradient of $AD = \dfrac{(18-15)}{(0-6)} = -\frac{1}{2}$;

equation of l is $y - 18 = -\frac{1}{2}(x - 0)$ i.e. $y = -\frac{1}{2}x + 18$.

(c) Since C lies on the curve, $16 = 16.4 - 16k \rightarrow k = 3$.

(d) B lies on both the line and the curve hence, $-\frac{1}{2}x + 18 = 16x - 3x^2$

$$6x^2 - 33x + 36 = 0$$
$$2x^2 - 11x + 12 = 0$$
$$(x - 4)(2x - 3) = 0$$
$$x = 4 \text{ or } x = \frac{3}{2}$$

When $x = \frac{3}{2}$, $y = 17\frac{1}{4}$, B is $\left(\frac{3}{2}, 17\frac{1}{4}\right)$.

> **Examiner's tips**
>
> Note that $(x - 4)$ must be a factor as C has x-coordinate of 4.

1.3 Sequences and series

(1) (a)

$$a + 2d = -20$$
$$a + 10d = 20$$

subtracting, $\quad 8d = 40 \rightarrow d = 5 \rightarrow a = -30$.

> nth term $= a + (n-1)d$ – you need to learn this!

(b)

$$\frac{k}{2}\{2a + (k - 1)d\} = 0$$
$$\frac{k}{2}\{-60 + 5(k - 1)\} = 0$$
$$\frac{k}{2}(5k - 65) = 0$$
$$5k - 65 = 0 \text{ since } k \neq 0$$
$$k = 13.$$

> You need to know this.
> Substituting for a and d.
> Do not multiply out!

(2) (a) The repayments form an AP with $a = 20$ and $d = 2$

$$n\text{th term} = a + (n - 1)d = 114$$
$$20 + (n - 1)2 = 114$$
$$2(n - 1) = 94$$
$$(n - 1) = 47$$
$$n = 48.$$

> Now make n the subject.

(b) The price of the car is the sum of the 48 repayments as the loan is interest-free

i.e. $P = S_{48} = \dfrac{(20+114)}{2} \times 48$

$\qquad = 67 \times 48 = £3,216.$

> Here it is easier to use $S_n = \dfrac{(a+l)n}{2}$.

(3) (a) $u_1 = 1$, $u_2 = \frac{1}{3}$, $u_3 = -\frac{1}{9}$.

(b) $\displaystyle\sum_{n=1}^{15} 3\left(\frac{2}{3}\right)^n = 2 + \frac{4}{3} + \frac{8}{9} + \dots$

> This is a GP with common ratio $\frac{2}{3}$ to 15 terms.

$\qquad = 2\left(1 - \left(\frac{2}{3}\right)^{15}\right)/\left(1 - \frac{2}{3}\right)$

$\qquad = 6\left(1 - \left(\frac{2}{3}\right)^{15}\right) = 5.986$ (4 s.f.)

> You need to know $S_n = a(1 - r^n)/(1 - r)$.

$\displaystyle\sum_{n=1}^{15} u_n = \sum_{n=1}^{15}\left\{3\left(\frac{2}{3}\right)^n - 1\right\} = \sum_{n=1}^{15} 3\left(\frac{2}{3}\right)^n - 15 = 5.986 - 15 = -9.014$ (4 s.f.).

(c) $LHS = 3\left\{3\left(\frac{2}{3}\right)^{n+1} - 1\right\} = 3^2\left(2^{n+1}\frac{1}{3}^{n+1}\right) - 3 = 3^{1-n}\,2^{n+1} - 3$

$RHS = 2\left\{3\left(\frac{2}{3}\right)^{n} - 1\right\} - 1 = 3^{1-n}\,2^{n+1} - 2 - 1 = 3^{1-n}\,2^{n+1} - 3 = LHS.$

(4) (a) $a(2)^2 = 15 \to 4a = 15 \to a = \frac{15}{4}$

6th term $= \frac{15}{4}.\,2^5 = \frac{15}{4}.\,32 = 120.$ ◄——— **Examiner's tips** *nth term $= ar^{n-1}$ – you need to learn this!*

(b) $S_{10} = \frac{15}{4}\,(2^{10} - 1)/(2 - 1)$ ◄——— *Use the form $Sn = a(r^n - 1)/(r - 1)$, since $r > 1$.*

$= 3836.25.$

(c) 2nd term $= \frac{15}{4} \times 2 = 7.5$; 5th term $= \frac{15}{4} \times 2^4 = 60$

hence first and second terms of AP are 7.5 and 60 i.e. $a = 7.5$ and $d = 52.5$

so 9th term $= a + 8d = 7.5 + 8 \times 52.5$ ◄——— *nth term $= a + (n - 1)d$ – you need to learn this!*

$= 427.5.$

(d) 13th term $= a + 12d = 7.5 + 12 \times 52.5 = 637.5$

$S_{13} = \frac{(7.5 + 637.5)}{2} \times 13$ ◄——— *Here we're using $Sn = \frac{(a + \ell)n}{2}$.*

$= 4192.5.$

(5) (a) $(x + 1)/(x + 4) = x/(x + 1),$ ◄——— *Since the ratios of successive terms are equal.*

$\to \quad (x + 1)^2 = x(x + 4)$ ◄——— *Cross-multiplying.*

$\to \quad x^2 + 2x + 1 = x^2 + 4x$

$\to \qquad 2x + 1 = 4x$

$\to \qquad x = \frac{1}{2}.$

(b) $r = x/(x + 1) = \frac{1}{2}/\left(\frac{1}{2} + 1\right) = \frac{1}{3}.$

(c) $S = a/(1 - r)$ ◄——— *You need to learn this result for the sum to infinity of a GP but note that it only applies for $-1 < r < 1$.*

and $a = x + 4 = 4.5$

$= 4.5/\left(1 - \frac{1}{3}\right)$

$= 6.75.$

(6) (a) After 1 month, the initial payment of £30 has a value of £30 + £30 $\frac{p}{100}$

$= £30\left(1 + \frac{p}{100}\right)$

$= £30r$

Hence, after 2 months, the initial payment of £30 has a value of £30$r \times r$

$= £30r^2$

Hence, after 12 months, the initial payment of £30 has a value of £30r^{12}.

(b) after 12 months, the 2nd payment of £30 has a value of £30r^{11}

after 12 months, the 3rd payment of £30 has a value of £30r^{10}

etc, etc,

after 12 months, the 12th payment of £30 has a value of £30r

Hence, the total value of all 12 payments

$= £\{30r^{12} + 30r^{11} + 30r^{10} + \ldots\ldots\ldots + 30\ r\}$

$= £\{30r + 30r^2 + 30r^3 + \ldots\ldots\ldots + 30r^{12}\}$

i.e. a GP with 12 terms

$= £30r\,(r^{12} - 1)/(r - 1).$ ◄——— *Using $Sn = a(r^n - 1)/(r - 1)$.*

Answers

(c) Now $p = \frac{1}{2}$ so $r = 1 + \frac{p}{100} = 1.005$

hence, the total value of all 12 payments
$$= £30 \times 1.005\,((1.005)^{12} - 1)/(1.005 - 1)$$
$$= £371.92 \text{ (to nearest penny)}$$

so, total interest $= £371.92 - (12 \times £30) = £11.92$.

Examiner's tips

(7)
$$\sum_{r=1}^{10} (r + 3^r) = \sum_{r=1}^{10} r + \sum_{r=1}^{10} 3^r$$
$$= \frac{(1+10)}{2} \times 10 + 3\,\frac{(3^{10} - 1)}{(3 - 1)}$$
$$= 55 + 88{,}572 = 88{,}627.$$

The first series is an AP, with $a = 1$ and $d = 1$ and $n = 10$.
The second series is a GP, with $a = 3$ and $r = 3$ and $n = 10$.

1.4 Trigonometry

(1)
$$3(1 - \sin^2 x) - 2\sin x = 2$$
$$3\sin^2 x + 2\sin x - 1 = 0$$
$$(3\sin x - 1)(\sin x + 1) = 0$$
$$\sin x = \tfrac{1}{3} \text{ or } \sin x = -1$$
$$x = 19.5°, 160.5° \text{ or } x = 270°.$$

Use $\sin^2 x + \cos^2 x = 1$ to give a quadratic in $\sin x$.
Collect all the terms on one side so that the $\sin^2 x$ term has a positive coefficient.
Factorise or use the quadratic formula.
Sin is positive in quadrants one and two.

(2) (a) $-1 = 3 + k\sin\frac{\pi}{2}$
$\rightarrow k = -4.$

Substituting for x and y.

(b) so $y = 3 - 4\sin x$
hence $y_{max} = 7.$

Greatest value of y will occur when $\sin x = -1$.

(3) (a) $\cos 3x = 0.5$ where $0 \leqslant 3x \leqslant 3\pi$
$3x = \frac{\pi}{3}, \frac{5\pi}{3}, \frac{7\pi}{3}$
$x = \frac{\pi}{9}, \frac{5\pi}{9}, \frac{7\pi}{9}.$

(b) $\tan\left(x + \frac{\pi}{2}\right) = -1$ where $\frac{\pi}{2} \leqslant \left(x + \frac{\pi}{2}\right) \leqslant \frac{3\pi}{2}$
$x + \frac{\pi}{2} = \frac{3\pi}{4}, \frac{7\pi}{4}$
hence $x = \frac{\pi}{4}$ is the only solution in the given range.

This value is out of range.

(4) (a) (i) Draw a right-angled triangle as follows:
Now find the missing adjacent side by using Pythagoras:

$$a^2 = 4^2 - \left(\sqrt{6} - \sqrt{2}\right)^2 = 16 - \left(6 - 2\sqrt{6}\sqrt{2} + 2\right)$$
$$= 8 + 2\sqrt{12}$$
$$= 8 + 4\sqrt{3}$$
$$a = \sqrt{\left(8 + 4\sqrt{3}\right)} \text{ so}$$
$$\cos 15° = \frac{\sqrt{\left(8 + 4\sqrt{3}\right)}}{4} = \sqrt{\left(8 + 4\sqrt{3}\right)/16} = \sqrt{\frac{1}{2} + \frac{1\sqrt{3}}{4}}.$$

Multiplying out the brackets.
Since $\sqrt{12} = \sqrt{4 \times 3} = 2\sqrt{3}$.

(ii) $\sin 105° = \sin 75° = \cos 15° = \sqrt{\frac{1}{2} + \frac{1\sqrt{3}}{4}}.$

Since $\sin x = \sin(180° - x)$ and $\cos 15° = \sin 75°$ from the triangle.

(b) $3\cos^2 x + \cos x - 2 = 0$
$(3\cos x - 2)(\cos x + 1) = 0$ ⟵
$\cos x = \frac{2}{3}$ or $\cos x = -1$
$x = 0.84$ or $(2\pi - 0.84) = 5.44$ or $x = \pi = 3.14$ (2 d.p.)
i.e. $x = 0.84, 5.44$ or 3.14.

Factorise or use the formula.

(5) (a) $120° = \frac{2\pi}{3}$ ⟵
Perimeter $= AB + $ arc AB
$\qquad = 2x\ 8\sin 60° + 8x\ \frac{2\pi}{3}$ ⟵
$\qquad = 8\sqrt{3} + 16\ \frac{\pi}{3} = 30.6$ cm (3 s.f.).

We need to use $s = r\theta$ where θ is in radians.

$\triangle OAB$ is isosceles so the perpendicular from O to AB meets AB at its mid-point, N, say where $AB = 2AN = 2x\ 8\sin60$.

(b)   Area = Area of sector OAB − Area $\triangle OAB$
$\qquad = \frac{1}{2}8^2\ \frac{2\pi}{3} - \frac{1}{2}8^2 \sin\frac{2\pi}{3}$ ⟵
$\qquad = 39.3$ cm^2.

Area of a sector is $\frac{1}{2}r^2\theta$ and area of $\triangle$ is $\frac{1}{2}ab\sin C$.

1.5 Differentiation

(1) $dy/dx = 3x^2 - 18x + 24$
For turning points, $3x^2 - 18x + 24 = 0$
$\qquad\qquad x^2 - 6x + 8 = 0$
$\qquad\qquad (x - 4)(x - 2) = 0$
$\qquad\qquad x = 4$ or   2
$\qquad\qquad y = 16$ or   20
$d^2y/dx^2 = 6x - 18$
When $x = 4$, $d^2y/dx^2 = 6$; when $x = 2$, $d^2y/dx^2 = -6$
The turning points are (4, 16) (minimum) and (2, 20) (maximum).

The gradient is zero at a turning point.
Simplify before solving.
We need both coordinates.
To prove that they are turning points we must show that $d^2y/dx^2 \neq 0$ at each point.

(2) (a)

It is always a good idea to draw a diagram.

Let h cm be the height of the tank. ⟵

A third variable will always be needed; this may sometimes be defined in the question itself.

Then the volume $hx^2 = 500$ litres $= 500,000$ cm^3. ⟵
Thus $h = \dfrac{500,000}{x^2}$ ⟵

$A = x^2 + 4hx$ ⟵
$\quad = x^2 + 4x\left(\dfrac{500,000}{x^2}\right)$ ⟵
$\quad = x^2 + \dfrac{2,000,000}{x}$. ⟵

We need the volume in cm^3 as h and x are in cm.
The constraint (here the volume being 500 litres) is used to obtain a connection between the two variables.
The base plus the four sides.
Substituting for h.
Cancelling the x.

(b) $A = x^2 + 2,000,000x^{-1}$ ⟵
$dA/dx = 2x - 2,000,000x^{-2}$ ⟵
Hence, $2x - 2,000,000x^{-2} = 0$
$x^3 - 1,000,000 = 0$
i.e. $x = 100$ ⟵
so $A = 100^2 + 2,000,000\ (100^{-1})$
$\qquad = 30,000$ is the minimum value of A.

All x terms need to be written as powers before differentiating.
Note that $-1 - 1 = -2$!
Dividing by 2 and multiplying through by x^{-2}, to clear the fractions.

Answers

(c) $d^2A/dx^2 = 2 + 4{,}000{,}000x^{-3}$
When $x = 100$, $d^2A/dx^2 = 6 > 0$, hence it is a minimum.

(3) (a) $dy/dx = 2 - 2x$
When $x = 0$, gradient of tangent is 2
$\rightarrow$ gradient of normal is $-\frac{1}{2}$

equation of normal at O is $y = -\frac{1}{2}x$.

Change the sign and invert to obtain the gradient of normal.

(b) At intersections of normal and curve,
$-\frac{1}{2}x = 2x - x^2$ i.e. $0 = 5x - 2x^2$
$0 = x(5 - 2x)$
$x = 0$ or $\frac{5}{2}$

Solving the normal and curve equations simultaneously.

When $x = \frac{5}{2}$, $y = 5 - \left(\frac{5}{2}\right)^2 = \frac{-5}{4}$
The point is $\left(\frac{5}{2}, \frac{-5}{4}\right)$.

(4) (a) $\qquad x + \frac{1}{4x} = -\frac{5}{4}$
$\qquad 4x^2 + 1 = -5x$
$\quad 4x^2 + 5x + 1 = 0$
$\quad (4x + 1)(x + 1) = 0$
$\quad x = -\frac{1}{4}$ or -1.

Multiplying through by $4x$ to clear the fractions.

(b) $f(x) = x + \left(\frac{1}{4}\right)x^{-1}$
$f'(x) = 1 - \left(\frac{1}{4}\right)x^{-2}$
$1 - \left(\frac{1}{4}\right)x^{-2} > 0$
$4x^2 - 1 > 0$
$(2x + 1)(2x - 1) > 0$

Before differentiating, write all terms as powers of x.

f is increasing if its gradient is positive. Multiplying through by $4x^2$ to clear the fractions.

To solve a quadratic inequality sketch its graph.

$x < -\frac{1}{2}$ or $x > \frac{1}{2}$.

1.6 Integration

(1) (a) $\dfrac{dy}{dx} = 13.5x^{-0.5} - 1.5x^{0.5}$
$\qquad = 1.5x^{-0.5}(9 - x).$

(b) $1.5x^{-0.5}(9 - x) = 0 \rightarrow x = 9$
When $x = 9$, $y = 27.(9)^{0.5} - 9^{1.5}$
$\qquad = 81 - 27 = 54$ i.e. A is (9, 54).

At a turning point the gradient is zero.
Both coordinates are required.

(c) Area $= \displaystyle\int_0^{27} 27x^{0.5} - x^{1.5} \, dx$

Area $= \displaystyle\int_a^b y \, dx$.

$\qquad = [18x^{1.5} - 0.4x^{2.5}]_0^{27}$
$\qquad = 1010$ (3 s.f.).

(2) (a) P is (2, 0); Q is (4, 0).

(b) $(x - 2) = (x - 2)(x - 4)$
$0 = (x - 2)(x - 4 - 1)$
$0 = (x - 2)(x - 5)$
$x = 2$ or $x = 5$
When $x = 5$, $y = 3$ i.e. R is (5, 3).

(c) Area $= \int_2^5 (x - 2) - (x - 2)(x - 4)\mathrm{d}x$

$= \int_2^5 (x - 2)(5 - x)\mathrm{d}x$

$= \int_2^5 - x^2 + 7x - 10\mathrm{d}x$

$= \left[-\tfrac{1}{3}x^3 + \tfrac{7}{2}x^2 - 10x\right]_2^5$
$= 4.5.$

(3) (a) $\mathrm{d}y/\mathrm{d}x = -2x$
When $x = 1$, $\mathrm{d}y/\mathrm{d}x = -2$
equation of tangent is $y - 3 = -2(x - 1)$
i.e. $y = -2x + 5$
$0 = -2x + 5 \rightarrow x = 2.5.$

(b) The curve crosses the x-axis when $0 = 4 - x^2$
i.e. $x = 2$ (or -2)
Shaded area $=$ Area under tangent between $x = 1$ and $x = 2.5$ $-$ Area under curve between $x = 1$ and $x = 2$

$= \tfrac{1}{2} \times 1.5 \times 3 - \int_1^2 4 - x^2\mathrm{d}x$

$= \tfrac{9}{4} - \left[4x - \tfrac{1}{3}x^3\right]_1^2$

$= \tfrac{9}{4} - \tfrac{5}{3}$

$= \tfrac{7}{12}.$

(4) (a) $\mathrm{d}y/\mathrm{d}x = 2(x - 2\sqrt{x} + 1)$
$= 2(x - 2x^{0.5} + 1)$
$y = \int 2x - 4x^{0.5} + 2\mathrm{d}x$
$= x^2 - \tfrac{8}{3}x^{1.5} + 2x + c.$

(b) $2 = 81 - 72 + 18 + c$
$\rightarrow c = -25$
so $y = x^2 - \tfrac{8}{3}x^{1.5} + 2x - 25.$

Questions with model answers

For help see Revise AS Study Guide pages 51 and 52

C grade candidate – mark scored 8/15

(1) **(a)** Find the coefficient of x^2 in the expansion of $(3 - 2x)^5$. **[4]**

Term is $\binom{5}{2} -2x^2\, 3^3$.

Hence coeff. $= 10 \times -2 \times 27 = -540$

(b) Find the constant term in the expansion of $(x^2 - \frac{1}{x})^6$. **[4]**

Term is $\binom{6}{2} (x^2)^2.(-\frac{1}{x})^4$

i.e. $\frac{6 \times 5}{1 \times 2} = 15$.

Given that the coefficients of x, x^2 and x^3 in the expansion of $(1 + x)^n$, where $n \geqslant 3$, are in arithmetic progression,

(c) find the value of n. **[7]**

coefficients are : $\binom{n}{1}, \binom{n}{2}, \binom{n}{3}$

so, $\binom{n}{3} - \binom{n}{2} = \binom{n}{2} - \binom{n}{1}$

N.B.

$n(n - 1)(n - 2)/6 - n(n - 1)/2 = n(n - 1)/2 - n$

$(n - 1)(n - 2) - 3(n - 1) = 3(n - 1) - 6$

$n^2 - 9n + 14 = 0$

$(n - 7)(n - 2) = 0$

$n = 7$ (as $n = 2$ is not possible since $n \geqslant 3$)

Examiner's Commentary

The candidate has forgotten to put brackets around –2x; forgetting to square the –2 is the most common error at this stage.

The binomial coefficient can be evaluated using the nCr button on a calculator, $\frac{5!}{3!\,2!}$, or more simply $\frac{5 \times 4}{1 \times 2}$, 2/4 scored.

This ensures that the x's cancel – the brackets are remembered here! 4/4 scored.

Correct but the candidate has always relied on the calculator to evaluate binomial coefficients, cannot cope with algebraic forms and stops, 2/7 scored.

Dividing by $n (\neq 0)$ and multiplying by 6 to clear fractions.

Multiplying out and collecting all terms on one side.

It is important that a reason is given for ignoring an answer.

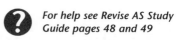

For help see Revise AS Study Guide pages 48 and 49

A grade candidate – mark scored 10/12

(2) The functions f, g are defined by

$$f: x \to 6x - 1, \; x \in \mathbb{R}$$

$$g: x \to \frac{4}{x-1}, \; x \in \mathbb{R}, \; x \neq 1.$$

(a) Find, in its simplest form, g^{-1}. **[4]**

Let $y = \dfrac{4}{x-1}$

$y(x-1) = 4$

$yx - y = 4$

$yx = 4 + y$

$x = \dfrac{4}{y} + 1$

so, $g^{-1}: x \to \dfrac{4}{x} + 1.$

(b) State the range of g. **[1]**

$y \geqslant 4.$

(c) Find, in its simplest form, $gf(x)$. **[2]**

$gf(x) = g\{f(x)\}$

$= \dfrac{4}{f(x)-1}$

$= \dfrac{4}{6x-2}.$

(d) Find the values of x for which $g(x) = f(x)$. **[5]**

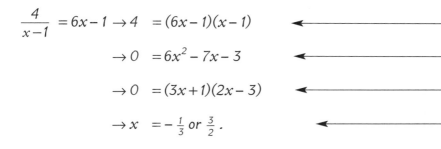

$\dfrac{4}{x-1} = 6x - 1 \to 4 = (6x-1)(x-1)$

$\to 0 = 6x^2 - 7x - 3$

$\to 0 = (3x+1)(2x-3)$

$\to x = -\dfrac{1}{3}$ or $\dfrac{3}{2}.$

Examiner's Commentary

The correct start – let $y = g(x)$ as this is equivalent to $g^{-1}(y) = x$ (or put $x = g(y)$, equivalent to $g^{-1}(x) = y$, and make y the subject).

Clearing the fractions.

Making x the subject – the RHS is now $g^{-1}(y)$, from above.

This is a correct definition but the domain, $x \in \mathbb{R}$, $x \neq 0$, has been omitted – a common error, losing a mark, **3/4 scored**.

Incorrect; generally it is much easier to find a domain; the range of g is the same as the domain of g^{-1} so the correct answer is $y \in \mathbb{R}$, $y \neq 0$, from part **(a)**. Note also that range of g^{-1} is same as domain of g, **0 scored**.

Putting this step in makes it obvious that gf means 'f then g'.

All correct, 2/2 scored.

Clearing the fractions.

Multiplying out and collecting all the terms on one side.

Refactorising.

All correct, **5/5 scored**.

Exam practice questions

 Answers on pp. 27–32

2.1 Algebra and functions

(1) Solve the equation $\dfrac{1}{x^2-4}+\dfrac{3}{x+2}=\dfrac{4}{5}$. [9]

(2) (a) Sketch on the same axes the graphs of $y=|x|$ and $y=2|x-1|$. [2]

(b) Solve the equation $|x|=2|x-1|$. [5]

(3) Sketch, on different axes, the graphs of

(a) $y=|\sin x|$, for $0° \leqslant x \leqslant 360°$. [2]

(b) $y=\sin|x|$, for $-360° \leqslant x \leqslant 360°$. [2]

(c) $y=\sin 2x$, for $0° \leqslant x \leqslant 360°$. [2]

(4) (a) Expand $(a+b)^4$. [4]

(b) Simplify $x^4+4x^3(1-x)+6x^2(1-x)^2+4x(1-x)^3+(1-x)^4$. [4]

(5) Prove that the equation $\dfrac{2x+3}{x-1}+\dfrac{3x-4}{x+1}=3$ has no real solutions. [7]

(6) The equation of a curve is $y=ax^n$, where a and n are constants. The points (2, 9) and (4, 15) both lie on the curve. Find the values of a and n. [9]

(7) Solve

(a) $\ln(a+10)=2\ln(a-2)$, [7]

(b) $2^{2x}-6.2^x+8=0$. [6]

2.2 Trigonometry

(1) Prove the following identities.

(a) $\left(\dfrac{1+\cos\theta}{\sin\theta}\right)^2+\left(\dfrac{1-\cos\theta}{\sin\theta}\right)^2=4\operatorname{cosec}^2\theta-2$, [6]

(b) $(\operatorname{cosec}x+\cot x)(\sec x+\tan x)=(1+\operatorname{cosec}x)(1+\sec x)$. [5]

(2) (a) Express $12\sin\theta+5\cos\theta$ in the form $R\sin(\theta+\alpha)$,
where $R>0$ and $0°<\alpha<90°$. [7]

(b) Find the maximum and minimum values of $(12\sin\theta+5\cos\theta)^3$, stating the values of θ at which they occur. [3]

(c) Solve, for $0°<\theta<360°$,
$$12\tan\theta+5=4\sec\theta.$$ [6]

(3) (a) Find all the values of x in the range $0 \leqslant x \leqslant 2\pi$, for which $\sin 3x = \sin x$. **[7]**

(b) Solve, for $0° < \theta < 360°$,
$$2\cos\theta + 7\cos\frac{\theta}{2} = 0.$$
[8]

2.3 and 2.4 Differentiation and integration

(1) (a) Sketch the curve with equation $y = 2e^x - 1$. **[2]**

(b) Find an equation of the normal to the curve at the point where $x = 0$. **[7]**

(2) $$f(x) = x + \frac{4}{x}, \ x \in \mathbb{R}, \ x > 0$$

(a) Find the coordinates of the turning point on the graph of $y = f(x)$. **[6]**

(b) Sketch the graph of $y = f(x)$. **[2]**

The region R is bounded by the curve $y = f(x)$, the lines $x = 1$ and $x = 3$ and the x-axis.

(c) Show, by shading, the region R on your graph. **[1]**

(d) Find the area of R. **[5]**

(e) Find the volume generated when R is rotated through 2π about the x-axis. **[6]**

(3)

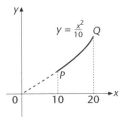

In the figure PQ is a sketch of the curve with equation $y = \frac{x^2}{10}$. The curved surface of an open bowl with a flat circular base is traced out when the arc PQ is rotated through one revolution about the y-axis. The radius of the base of the bowl is 10 cm and the radius of the top rim of the bowl is 20 cm.

(a) Find the volume of the bowl, in litres to 1 decimal place. **[9]**

The point R lies on the arc PQ. The curved surface of another bowl is traced out when the arc PR is rotated through one revolution about the y-axis.
The volume of this bowl is 10 litres.

(b) Find the depth of this bowl in cm, to 1 decimal place. **[8]**

Exam practice questions

(4)

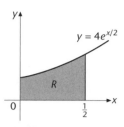

In the diagram, the shaded region, R, is bounded by the curve with equation $y = 4e^{x/2}$, the line with equation $x = \frac{1}{2}$ and the x and y axes. The region R is rotated through 360° about the x-axis. Find, in terms of e and π, the volume of the solid generated. **[5]**

(5) Find

(a) $\int x^{-1/4} \, dx.$ **[1]**

(b) $\int \left(x - \frac{3}{x}\right)^2 \, dx.$ **[4]**

(c) $\int \frac{1}{2}e^x \, dx.$ **[1]**

(6) **(a)** Sketch the graph of $y = 2 - 8x^2$, indicating clearly any points of intersection with the coordinate axes. **[4]**

(b) Find an equation of the tangent to this curve at the point where $x = p$. **[6]**

(c) Find the value of p for which this tangent is perpendicular to the line with equation $y = 2x + 1$. **[3]**

The region R is bounded by the curve and the x-axis.

(d) Find the area of R. **[5]**

The region R is rotated through 2π about the y-axis.

(e) Calculate the volume of the solid formed, leaving your answer in terms of π. **[5]**

(7) **(a)** Sketch the curve with equation $y = \ln x$. **[1]**

(b) Show that the equation
$$x + \ln x - 2 = 0$$
has one real root, α, and that $1.5 < \alpha < 2$. **[4]**

(8) **(a)** Show graphically that the equation $x^3 - 8e^{-x} = 0$ has one real root, α. **[4]**

(b) Find the integer n such that $n < \alpha < n + 1$. **[4]**

(c) Use the iteration formula
$$x_{n+1} = (8e^{-x}{}_n)^{1/3}, \; x_1 = 2$$
to find α correct to 1 decimal place. **[4]**

(d) Prove that your answer is the value of α to 1 decimal place. **[3]**

Answers

2.1 Algebra and functions

(1) $\dfrac{1}{x^2 - 4} + \dfrac{3}{x + 2} = \dfrac{4}{5} \rightarrow \dfrac{1}{(x+2)(x-2)} + \dfrac{3}{x+2} = \dfrac{4}{5}$

Factorise the denominator.
Now multiply through by the lowest common denominator, $5(x+2)(x-2)$.

$\rightarrow 5 + 15(x - 2) = 4(x + 2)(x - 2)$ *This clears all the fractions.*
$\rightarrow 15x - 25 = 4(x^2 - 4)$
$\rightarrow 4x^2 - 15x + 9 = 0$ *Multiplying out and collecting terms.*
$\rightarrow (4x - 3)(x - 3) = 0$ *Factorising.*
$\rightarrow x = \frac{3}{4}$ or $x = 3$.

(2) (a)

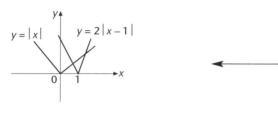

Draw $y = x$ and $y = 2(x - 1)$ and reflect the parts of the graphs below the x-axis in the x-axis.

(b) $|x| = 2|x - 1|$; from the graph there are two cases:
if $x < 1$, $x = 2(1 - x) \rightarrow 3x = 2 \rightarrow x = \frac{2}{3}$
if $x > 1$, $x = 2(x - 1) \rightarrow x = 2$.

$2|x - 1| = 2(x - 1)$ if $x > 1$ and $2(1 - x)$ if $x \leqslant 1$.

(3) (a)

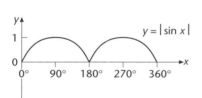

Draw the graph of $y = \sin x$ and reflect the parts of the graph below the x-axis in the x-axis.

(b)

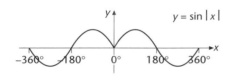

Draw the graph of $y = \sin x$ for $x \geqslant 0$ and reflect this in the y-axis.

(c)

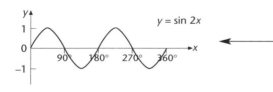

To obtain the required graph, stretch the graph of $y = \sin x$ by scale factor $\frac{1}{2}$, parallel to the x-axis.

(4) (a) $(a + b)^4 = a^4 + 4a^3 b + 6a^2 b^2 + 4ab^3 + b^4$.

In each term the sum of the powers of a and b must be 4; the coefficients come from Pascal's Triangle.

Answers

(b) Putting $a = x$ and $b = (1 - x)$ on the RHS gives the expression requiring simplification. Hence, from part **(a)**, it simplifies to $(x + 1 - x)^4 = 1^4 = 1$.

A much longer method would be to expand all the brackets and then collect terms! It is always worth a few seconds thought before launching into a routine method.

(5) $\dfrac{2x+3}{x-1} + \dfrac{3x-4}{x+1} = 3$

$\to (2x + 3)(x + 1) + (3x - 4)(x - 1) = 3(x - 1)(x + 1)$

$\to 2x^2 + 5x + 3 + 3x^2 - 7x + 4 = 3x^2 - 3$

$\to 2x^2 - 2x + 10 = 0$

'$b^2 - 4ac$' $= 4 - 80 < 0$ so no real roots.

Now multiply through by the lowest common denominator, $(x + 1)(x - 1)$. This clears all the fractions.

Multiplying out and collecting terms.

(6) $15 = a4^n$

$9 = a2^n$

dividing gives

$\dfrac{5}{3} = 2^n$

$\log\dfrac{5}{3} = \log 2^n$

$\log\dfrac{5}{3} = n\log 2$

$\dfrac{\log\frac{5}{3}}{\log 2} = n$, i.e. $n = 0.737$ (3 s.f.)

hence, $9 = \dfrac{5a}{3}$

$a = \dfrac{27}{5} = 5.4$.

Since both points lie on the curve.

To eliminate a.

Note that $4^n/2^n = \left(\frac{4}{2}\right)^n = 2^n$.

Generally if the unknown is a power we take logs.in 2^x.

Since $\frac{5}{3} = 2^n$.

(7) (a) $\ln(a + 10) = 2\ln(a - 2)$

$\ln(a + 10) = \ln(a - 2)^2$

$(a + 10) = (a - 2)^2 = a^2 - 4a + 4$

$a^2 - 5a - 6 = 0$

$(a - 6)(a + 1) = 0$

$a = 6$ or $a = -1$

but $a > 2$, so $a = 6$.

Note $a > 2$ as we cannot take the log of a negative number.

Using the 3rd law of logarithms.

(b) $2^{2x} - 6.2^x + 8 = 0$

$(2^x)^2 - 6.2^x + 8 = 0$

$(2^x - 4)(2^x - 2) = 0$

$2^x = 4$ or $2^x = 2$

$x = 2$ or $x = 1$.

Taking logs here doesn't help since log 0 is undefined and we cannot simplify the logarithm of a sum or difference.

The key step – we now have a quadratic in 2^x.

2.2 Trigonometry

(1) (a) LHS $= \dfrac{1 + 2\cos\theta + \cos^2\theta + 1 - 2\cos\theta + \cos^2\theta}{\sin^2\theta}$

$= \dfrac{2 + 2(1 - \sin^2\theta)}{\sin^2\theta}$

$= 4\operatorname{cosec}^2\theta - 2 =$ RHS.

Squaring the brackets and using the common denominator.

(b) $\quad$ LHS $= \dfrac{(1+\cos x)}{\sin x} \dfrac{(1+\sin x)}{\cos x}$ $\longleftarrow$

Change everything into sines and cosines.

$\quad = \dfrac{(1+\cos x)(1+\sin x)}{\sin x \quad \cos x}$

$\quad = \dfrac{(1+\cos x)(1+\sin x)}{\cos x \quad \sin x}$

$\quad = (\sec x + 1)(\operatorname{cosec} x + 1) = $ RHS.

(2) (a) $\quad 12\sin\theta + 5\cos\theta = R\sin(\theta + \alpha)$

$\quad = R\sin\theta\cos\alpha + R\cos\theta\sin\alpha$ $\longleftarrow$ $\quad$ *Using the $\sin(A + B)$ formula.*

$\quad = (R\cos\alpha)\sin\theta + (R\sin\alpha)\cos\theta$

$\quad$ So $\quad 12 = R\cos\alpha$ $\longleftarrow$ $\quad$ *Since this is an identity we can equate*

$\quad$ And $\quad 5 = R\sin\alpha$ $\quad\quad\quad\quad\quad\quad\quad\quad$ *coefficients of $\sin\alpha$ and $\cos\alpha$.*

$\quad\quad \rightarrow \dfrac{5}{12} = \tan\alpha$ $\longleftarrow$ $\quad$ *Dividing the equations to eliminate R.*

$\quad\quad \rightarrow \alpha = 22.6°$ (3 s.f.)

$\quad$ And $\quad R^2 = 5^2 + 12^2$ $\longleftarrow$ $\quad$ *Squaring and adding the equations*

$\quad$ i.e. $\quad R = 13$ (as $R > 0$). $\quad\quad\quad\quad$ *$(\cos^2\alpha + \sin^2\alpha = 1)$.*

(b) $\quad$ We have $12\sin\theta + 5\cos\theta = 13\sin(\theta + \alpha)$

$\quad$ So $(12\sin\theta + 5\cos\theta)^3 = 13^3\sin^3(\theta + \alpha)$

$\quad$ Hence maximum value is 13^3 when $\theta + \alpha = 90°$ $\longleftarrow$ *Since maximum value of sin is 1.*

$\quad$ i.e. 2197 when $\theta = 67.4°$

$\quad$ Hence minimum value is $(-13)^3$ when $\theta + \alpha = 270°$ $\longleftarrow$ *Since minimum value of sin is -1.*

$\quad$ i.e. -2197 when $\theta = 247.4°$.

(c) $\quad 12\tan\theta + 5 = 4\sec\theta$

$\quad 12\sin\theta + 5\cos\theta = 4$ $\longleftarrow$ $\quad$ *Multiplying through by $\cos\theta$.*

$\quad 13\sin(\theta + \alpha) = 4$ $\longleftarrow$ $\quad$ *Using part (a).*

$\quad \sin(\theta + \alpha) = \dfrac{4}{13}$

$\quad \theta + 22.6° = 377.9°$ or $162.1°$ $\longleftarrow$ *$17.9°$ would give a θ value which is out*

$\quad \theta = 355.3°$ or $139.5°$. $\quad\quad\quad\quad$ *of range.*

(3) (a) $\quad\quad\quad\quad \sin 3x = \sin x$

$\quad \rightarrow \quad \sin 3x - \sin x = 0$

$\quad \rightarrow \quad 2\sin x \cos 2x = 0$ $\longleftarrow$ $\quad$ *Factorising using $\sin P - \sin Q$ formula.*

$\quad \rightarrow \quad \sin x = 0$ or $\cos 2x = 0$

$\quad \rightarrow \quad x = 0, \pi, 2\pi$ or $2x = \dfrac{\pi}{2}, \dfrac{3\pi}{2}, \dfrac{5\pi}{2}, \dfrac{7\pi}{2}$

$\quad\quad\quad\quad\quad\quad x = \dfrac{\pi}{4}, \dfrac{3\pi}{4}, \dfrac{5\pi}{4}, \dfrac{7\pi}{4}$.

(b) $\quad 2\cos\theta + 7\cos\dfrac{\theta}{2} = 0$

$\quad 2\left(2\cos^2\dfrac{\theta}{2} - 1\right) + 7\cos\dfrac{\theta}{2} = 0$ $\longleftarrow$ *Using $\cos 2A = 2\cos^2 A - 1$, with $A = \dfrac{\theta}{2}$.*

$\quad 4\cos^2\dfrac{\theta}{2} + 7\cos\dfrac{\theta}{2} - 2 = 0$

$\quad \left(4\cos\dfrac{\theta}{2} - 1\right)\left(\cos\dfrac{\theta}{2} + 2\right) = 0$

$\quad \left(4\cos\dfrac{\theta}{2} - 1\right) = 0$ or $\cos\dfrac{\theta}{2} + 2 = 0$

$\quad \cos\dfrac{\theta}{2} = \dfrac{1}{4}$ $\longleftarrow$ $\quad$ *$\cos\dfrac{\theta}{2} + 2 = 0$ is impossible.*

$\quad \dfrac{\theta}{2} = 75.5°$ $\longleftarrow$ $\quad$ *$284.5°$ would give a θ value which is*

$\quad \theta = 151°$. $\quad\quad\quad\quad\quad\quad$ *out of range.*

Answers

2.3 and 2.4 Differentiation and integration

(1) (a)

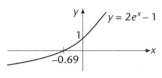

Show where the graph cuts the axes.
When $y = 0$,
$e^x = 0.5 \rightarrow x = \ln 0.5 = -0.69$.

(b) $y = 2e^x - 1$

$dy/dx = 2e^x$

When $x = 0$, $dy/dx = 2 \rightarrow$ gradient of normal is $-\frac{1}{2}$

When $x = 0$, $y = 1$

Equation of normal is $y - 1 = -\frac{1}{2}(x - 0)$

i.e. $2y + x - 2 = 0$.

Invert and change the sign to obtain the gradient of the normal.
Using $y - y_1 = m(x - x_1)$.

(2) (a) $y = x + \frac{4}{x} = x + 4x^{-1}$

$dy/dx = 1 - 4x^{-2}$

$0 = 1 - 4x^{-2} \rightarrow 0 = x^2 - 4$

So $x = 2$ (since $x > 0$)

$y = 2 + \frac{4}{2} = 4$

Turning point at $(2, 4)$.

Express each term as a power of x before differentiating

At turning point the gradient is zero; multiply through by x^2 to clear negative power.

(b)

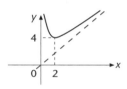

Only $x > 0$ is required.
For small x, graph is similar to $y = \frac{4}{x}$.
For large x, graph is similar to $y = x$
($y = x$ is an oblique asymptote).

(c)

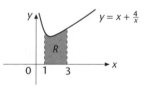

(d) Area $= \int_1^3 x + 4x^{-1} \, dx$

$= \left[\frac{1}{2}x^2 + 4\ln x \right]_1^3$

$= \frac{1}{2}(3^2 - 1^2) + 4(\ln 3 - \ln 1)$

$= 4 + 4\ln 3$.

Area $= \int_a^b y \, dx$.

(e) Volume $= \pi \int_1^3 (x + 4x^{-1})^2 \, dx = \pi \int_1^3 (x^2 + 8 + 16x^{-2}) \, dx$

$= \pi \left[\frac{1}{3}x^3 + 8x - \frac{16}{x} \right]_1^3$

$= \frac{106\pi}{3}$.

Volume $= \pi \int_a^b y^2 \, dx$.

(3) (a) At P, $x = 10 \rightarrow y = 10$

At Q, $x = 20 \rightarrow y = 40$

Volume $= \pi \int_{10}^{40} x^2 \, dy$

The y-coordinates are needed as we are rotating around the y-axis.

$= \pi \int_{10}^{40} 10y \, dy$

Use the equation of the curve to express x^2 in terms of y.

$= 5\pi \left[y^2 \right]_{10}^{40}$

$= 5\pi(40^2 - 10^2) = 7500\pi \text{ cm}^3$

$= 23.6 \text{ litres (1 d.p.).}$

$1000 \text{ cm}^3 = 1$ litre.

(b) $10\,000 = 5\pi \left[y^2 \right]_{10}^{r}$

Where r is the y-coordinate of R.

$10\,000 = 5\pi(r^2 - 10^2)$

Putting in the limits.

$\dfrac{2000}{\pi} = r^2 - 100$

$100 + \dfrac{2000}{\pi} = r^2$

Solving for r.

$r = 27.1 \text{ (1 d.p.)}$

Depth of bowl is $27.1 - 10 = 17.1$ cm.

Depth $= y_R - y_P$.

(4) Volume $= \pi \int_{0}^{1/2} y^2 \, dx$

Now use the equation of the curve to express y^2 in terms of x.

$= \pi \int_{0}^{1/2} 16e^x \, dx$

$(4e^{x/2})^2 = 16e^x$.

$= 16\pi \left[e^x \right]_{0}^{1/2}$

$= 16\pi(e^{0.5} - 1).$

(5) (a) $\int x^{-1/4} \, dx = \frac{4}{3} x^{3/4} + c.$

It is best to use top-heavy fractions.

(b) $\int (x - \frac{3}{x})^2 \, dx$

$= \int x^2 - 6 + \frac{9}{x^2} dx$

Multiply out brackets first.

$= \int x^2 - 6 + 9x^{-2} dx$

Express each term as a power of x before integrating.

$= \frac{1}{3}x^3 - 6x - 9x^{-1} + c.$

Always include an arbitrary constant.

(c) $\int \frac{1}{2} e^x \, dx = \frac{1}{2} e^x + c.$

(6) (a) When $x = 0$, $y = 2$

When $y = 0$, $2 - 8x^2 = 0 \rightarrow x = \pm\frac{1}{2}$

The negative coefficient of x^2 indicates that the parabola has a maximum point.

(b) $y = 2 - 8x^2$

$dy/dx = -16x$

Equation of tangent is

$y - (2 - 8p^2) = -16p \, (x - p)$

Using $y - y_1 = m(x - x_1)$.

i.e. $y + 16px = 8p^2 + 2.$

Examiner's tips

(c)   Gradient of $y = 2x + 1$ is 2

$\rightarrow -16p = -\frac{1}{2} \rightarrow p = \frac{1}{32}$.

Invert and change the sign to obtain the gradient of the perpendicular.

(d)   Area $= \int_{-\frac{1}{2}}^{\frac{1}{2}} y\,dx = \int_{-\frac{1}{2}}^{\frac{1}{2}} 2 - 8x^2\,dx$

$= 2\left[2x - \frac{8}{3}x^3\right]_0^{\frac{1}{2}}$

Using the symmetry of the graph.

$= \frac{4}{3}$.

(e)   Volume $= \pi \int_0^2 x^2\,dy$

$y = 2 - 8x^2 \rightarrow 8x^2 = 2 - y \rightarrow x^2 = \frac{1}{8}(2 - y)$.

$= \frac{\pi}{8} \int_0^2 2 - y\,dy$

$= \frac{\pi}{8}\left[2y - \frac{1}{2}y^2\right]_0^2$

$= \frac{\pi}{4}$.

(7) (a)

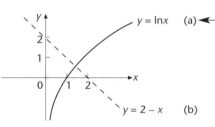

This graph is a reflection of $y = e^x$ in the line $y = x$ since $\ln x$ is the inverse of e^x; passing through (1, 0).

(b) $x + \ln x - 2 = 0 \rightarrow \ln x = 2 - x$

The solutions of this equation are the x-coordinates of the points of intersection of the graphs of $y = \ln x$ and $y = 2 - x$; so draw $y = 2 - x$ on the sketch above.

Clearly only one point of intersection.

Let $f(x) = x + \ln x - 2$; then $f(1.5) = -0.0945...$ and $f(2) = 0.69...$

The change of sign indicates that $1.5 < \alpha < 2$, since the graph of $y = f(x)$ is continuous.

(8) (a) $x^3 - 8e^{-x} = 0 \rightarrow x^3 = 8e^{-x}$

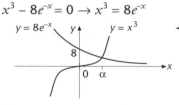

Draw the graphs of $y = x^3$ and $y = 8e^{-x}$ on the same axes.

Only one point of intersection at $x = \alpha$.

(b)   Let $f(x) = x^3 - 8e^{-x}$

$f(0) = -8$; $f(1) = -1.94...$; $f(2) = 6.91...$

The function changes sign between $x = 1$ and $x = 2$.

$\rightarrow 1 < \alpha < 2$; hence $n = 1$.

$x_2 = 1.0268...$

$x_3 = 1.42...$

$x_4 = 1.2457...$

$x_5 = 1.320...$

$x_6 = 1.287...$

$x_7 = 1.30...$ $\alpha = 1.3$.

It is reasonable to assume that $\alpha = 1.3$ (1 d.p.) but this is the only way to prove it.

(c) $f(1.25) = -0.33...$ and $f(1.35) = 0.386...$

$\rightarrow 1.25 < \alpha < 1.35$ i.e. $\alpha = 1.3$ (1 d.p.).

Questions with model answers

For help see Revise AS
Study Guide pages 72–4

C grade candidate – mark scored 9/15

Examiner's Commentary

(1) At 12 noon ship S has position vector $(-9\mathbf{i} + 6\mathbf{j})$ km and is moving with constant velocity $(3\mathbf{i} + 12\mathbf{j})$ km h^{-1} and ship T has position vector $(16\mathbf{i} + 6\mathbf{j})$ km and is moving with constant velocity $(-9\mathbf{i} + 3\mathbf{j})$ km h^{-1}.

(a) Find how far apart the ships are at 12 noon. **[4]**

$16\mathbf{i} + 6\mathbf{j} - -9\mathbf{i} + 6\mathbf{j} = 25\mathbf{i} + 12\mathbf{j}$

$Distance = \sqrt{(25^2 + 12^2)} = 27.7\ km$

Incorrect – the brackets around the second vector have been omitted resulting in the **j** terms being added instead of being subtracted, **2/4 scored**.

(b) Write down the position vectors of S and T at time t hours after noon and hence find the vector from S to T at time t hours after noon. **[5]**

$\mathbf{s} = (-9\mathbf{i} + 6\mathbf{j}) + t(3\mathbf{i} + 12\mathbf{j})$

Correct

$\mathbf{t} = (16\mathbf{i} + 6\mathbf{j}) + t(-9\mathbf{i} + 3\mathbf{j})$

Correct

$\overrightarrow{ST} = \mathbf{t} - \mathbf{s} = (16\mathbf{i} + 6\mathbf{j}) + t(-9\mathbf{i} + 3\mathbf{j}) - \{(-9\mathbf{i} + 6\mathbf{j}) + t(3\mathbf{i} + 12\mathbf{j})\}$

$= 25\mathbf{i} + t(-12\mathbf{i} - 9\mathbf{j})$

The brackets this time are remembered! Putting $t = 0$ in the answer would have revealed the error in part (a), **5/5 scored**.

(c) Find the least distance between the ships in the subsequent motion. **[6]**

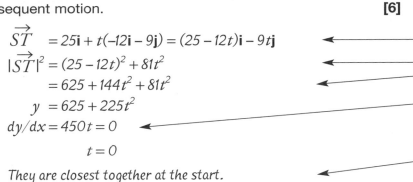

$\overrightarrow{ST} = 25\mathbf{i} + t(-12\mathbf{i} - 9\mathbf{j}) = (25 - 12t)\mathbf{i} - 9t\mathbf{j}$

$|\overrightarrow{ST}|^2 = (25 - 12t)^2 + 81t^2$

$= 625 + 144t^2 + 81t^2$

$y = 625 + 225t^2$

$dy/dx = 450t = 0$

$t = 0$

They are closest together at the start.

A good start – collecting the **i**'s and **j**'s.
Also correct.
Incorrect algebra – 600t has been missed out.
Wrong notation – not dy/dx but dy/dt – but the method is correct.

Correct conclusion from wrong working, **2/6 scored**.

Questions with model answers

? *For help see Revise AS Study Guide pages 77 and 80*

A grade candidate – mark scored 11/14

Examiner's Commentary

(2) A particle P of mass 2 kg is acted upon by two horizontal forces $(2\mathbf{i} + 3\mathbf{j})$N and $(4\mathbf{i} - 7\mathbf{j})$N where $\mathbf{i}$ and $\mathbf{j}$ are unit horizontal vectors due East and due North respectively.

Find

(a) the magnitude of the acceleration of P: **[4]**

$(2\mathbf{i} + 3\mathbf{j}) + (4\mathbf{i} - 7\mathbf{j}) = 2\mathbf{a}$

$3\mathbf{i} - 2\mathbf{j} = \mathbf{a}$

$|\mathbf{a}| = \sqrt{3^2 + (-2)^2} = \sqrt{13} = 3.61$

Magnitude of acceleration is 3.61 m s^{-2}

Applying $\mathbf{F} = m\mathbf{a}$.

Using Pythagoras.

All correct, 4/4 scored.

(b) the direction of the acceleration of P: **[3]**

A diagram is a good idea.

$\tan\theta = \frac{2}{3}$

$\theta = 33.7°$. *The direction of acceleration is 33.7° S of East.*

All correct, 3/3 scored.

At time $t = 0$, P is at the point with position vector $(\mathbf{i} - \mathbf{j})$ m and is moving with velocity $(\mathbf{i} + \mathbf{j})$ m s^{-1}.

Find, when $t = 4$ s,

(c) the speed of P: **[4]**

$\mathbf{v} = (\mathbf{i} + \mathbf{j}) + 4(3\mathbf{i} - 2\mathbf{j}) = 13\mathbf{i} - 7\mathbf{j}$

Using $\mathbf{v} = \mathbf{u} + \mathbf{a}t$, but the candidate forgets to find the speed, i.e. $|\mathbf{v}|$, 2/4 scored.

(d) the position vector of P: **[3]**

$\mathbf{r} = 4(\mathbf{i} + \mathbf{j}) + \frac{1}{2}(3\mathbf{i} - 2\mathbf{j})4^2$

$= 4\mathbf{i} + 4\mathbf{j} + 24\mathbf{i} - 16\mathbf{j}$

$= 28\mathbf{i} - 12\mathbf{j}$

Using $\mathbf{s} = \mathbf{u}t + \frac{1}{2}\mathbf{a}t^2$.

Correct so far but the candidate forgets to add his answer onto the initial position vector $(\mathbf{i} - \mathbf{j})$, 2/3 scored.

Exam practice questions

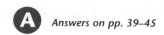

Answers on pp. 39–45

3.1 Vectors

(1) Three forces $\mathbf{F}_1$, $\mathbf{F}_2$ and $\mathbf{F}_3$ act on a particle.
$\mathbf{F}_1 = (2\mathbf{i} + 3a\mathbf{j})$N, $\mathbf{F}_2 = (a\mathbf{i} + b\mathbf{j})$N, $\mathbf{F}_3 = (b\mathbf{i} + 4\mathbf{j})$N
Given that the particle is in equilibrium, find the values of a and b. **[7]**

(2) Two forces, of magnitude P N and Q N, have a resultant of magnitude $2\sqrt{5}$ N when the angle between their lines of action is 90°. When the angle between their lines of action is 60° the magnitude of their resultant is $2\sqrt{7}$ N. Find the values of P and Q. **[12]**

(3) A particle is acted upon by two forces $\mathbf{F}_1$ and $\mathbf{F}_2$.
$$\mathbf{F}_1 = (2\mathbf{i} + \mathbf{j})\text{N and } \mathbf{F}_2 = (a\mathbf{i} + 2a\mathbf{j})\text{N}.$$

(a) Find the angle between $\mathbf{F}_1$ and $\mathbf{j}$. **[2]**

The resultant $\mathbf{R}$ of $\mathbf{F}_1$ and $\mathbf{F}_2$ is parallel to $\mathbf{i}$.

(b) Find the magnitude of $\mathbf{R}$. **[6]**

3.2 Kinematics

(Take $g = 9.8$ m s^{-2})

(1) A cricket ball is thrown vertically upwards from ground level and takes 5 s to reach the ground again. Find the maximum height of the ball above the ground. **[6]**

(2) A car moves with constant acceleration from a speed of 14 m s^{-1} to a speed of 34 m s^{-1} in 20 s.

(a) Find how far the car travels during this 20 s period. **[2]**

(b) Find how long it takes to cover half of this distance. **[7]**

(3) A small stone is dropped from the top of a tower. One second later another small stone is thrown vertically downwards from the same point at 19.6 m s^{-1}. Given that the two stones hit the ground at the same time, and assuming no air resistance, calculate the height of the tower. **[7]**

(4) A car travelling along a straight road takes two minutes to travel between two sets of traffic lights which are 2145 m apart. It starts at rest and accelerates uniformly for 30 s. It then moves with constant speed before uniformly decelerating to rest for the final 15 s.

(a) Illustrate the motion on a velocity–time graph. **[2]**

(b) Find the acceleration of the car. **[5]**

Exam practice questions

(5) Two cars A and B move along a straight horizontal road with constant acceleration. Car A has acceleration 2 m s^{-2} and B has acceleration 1 m s^{-2}. At time $t = 0$ car A has speed 1 m s^{-1} and is at the point O, and at $t = 4$ s car B is at O and has speed 16. Find

    (a)   the times between which car B is ahead of car A,   **[10]**

    (b)   the distance from O at which car B overtakes car A,   **[2]**

    (c)   the distance from O at which car A overtakes car B.   **[2]**

(6) A stone is projected with speed u m s^{-1} at an angle α above the horizontal from the top of a vertical cliff which is 56 m high. The stone moves in a vertical plane which is perpendicular to the cliff and lands in the sea 4 s later at a point which is 32 m from the foot of the cliff. Find

    (a)   the value of u,   **[4]**

    (b)   the value of α,   **[4]**

    (c)   the speed with which the stone hits the sea,   **[5]**

    (d)   the direction of motion of the stone when it hits the sea.   **[3]**

(7) A ball is projected with speed 49 m s^{-1} at an angle of 30° above the horizontal from the top of a vertical cliff which is 196 m high. The ball moves in a vertical plane which is perpendicular to the cliff and lands in the sea at the point P. Find

    (a)   the greatest height of the ball above the sea,   **[4]**

    (b)   the time taken to hit the sea,   **[7]**

    (c)   the distance of P from the foot of the cliff.   **[2]**

(8) A train travels in a straight line between two railway stations A and B, stopping at both. There is a signal box S between the two stations and the train passes S at exactly 12 noon. The velocity of the train, v km h^{-1}, at t minutes past noon is given by

$$v = \tfrac{15}{2}(3 + 4t - 4t^2).$$

Find

    (a)   the velocity of the train as it passes the signal box,   **[1]**

    (b)   the time of departure from A and arrival at B,   **[6]**

    (c)   the maximum velocity of the train,   **[5]**

    (d)   the average velocity of the train between A and B.   **[6]**

3.3 Statics

(Take $g = 9.8$ m s^{-2})

(1) A particle is suspended by two light inextensible strings and hangs in equilibrium. One string is inclined at 60° to the horizontal and the tension in that string is 20 N. The other string is inclined at 30° to the horizontal. Find, to 3 significant figures,

  (a) the weight of the particle, **[3]**

  (b) the tension in the second string. **[3]**

(2) A particle is placed on a rough plane inclined at an angle α to the horizontal, where $\tan\alpha = \frac{3}{4}$. The particle is kept in equilibrium by a horizontal force of magnitude 10 N, acting in the vertical plane containing the line of greatest slope of the inclined plane through the particle. The coefficient of friction between the particle and the plane is $\frac{1}{2}$. Given that the particle is on the point of slipping up the plane, find the normal reaction of the plane on the particle, and the weight of the particle. **[9]**

(3) A uniform beam AB of mass 10 kg and length 2.4 m is at rest in equilibrium in a horizontal position. The beam is supported by two vertical ropes attached to the beam at the points X and Y where $AX = 0.4$ m and $YB = 0.6$ m. Find the tension in each rope. **[6]**

(4) A non-uniform rod AB has length $3a$ and mass $5m$. It rests in equilibrium in a horizontal position on two supports at the points P and Q, where $AP = PQ = QB = a$. A particle of mass $2m$ is fixed to the rod at B. Given that the rod is on the point of tilting about Q, find the distance of the centre of mass of the rod from B. **[5]**

Exam practice questions

3.4 Dynamics

(Take $g = 9.8$ m s^{-2})

(1)

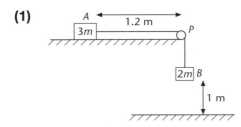

A particle A of mass $3m$ is placed on a rough horizontal table. The particle is attached to one end of a light inextensible string which passes over a small smooth fixed pulley P at the edge of the table. Another particle B of mass $2m$ is attached to the other end of the string and B hangs freely. AP is perpendicular to the edge of the table and A, P and B lie in the same vertical plane. The system is released from rest, with the string taut, when A is 1.2 m from the edge of the table and B is 1 m above the floor, as shown in the diagram.

Given that B strikes the floor after 2 s and does not rebound, find

(a)   the acceleration of A during the first two seconds of the motion,   **[2]**

(b)   to 2 decimal places, the coefficient of friction between A and the table,   **[10]**

(c)   by calculation, whether the particle A reaches the pulley.   **[6]**

(2)   A particle P of mass $2m$ is moving on a smooth horizontal plane with a speed u when it collides with another particle Q of mass km whose speed is $3u$ in the opposite direction. As a result of the collision the directions of motion of both particles is reversed and the speed of P is halved.

(a)   Find the range of values of k.   **[5]**

(b)   Find the magnitude of the impulse on P from Q.   **[3]**

(3)   A particle P of mass 2 kg is pushed, by a constant horizontal force of magnitude 30 N, up a rough plane inclined to the horizontal at an angle α, where $\tan\alpha = \frac{3}{4}$. The particle P moves with constant acceleration a m s^{-2}. The coefficient of friction between the particle and the inclined plane is 0.2.

(a)   Find the magnitude of the normal reaction between the particle and the inclined plane.   **[4]**

(b)   Find the value of a.   **[4]**

Answers

3.1 Vectors

(1)

$(2\mathbf{i} + 3a\mathbf{j}) + (a\mathbf{i} + b\mathbf{j}) + (b\mathbf{i} + 4\mathbf{j}) = 0$ ◄— *Since the forces are in equilibrium, their resultant is zero.*

$(2 + a + b)\,\mathbf{i} + (3a + b + 4\mathbf{j}) = 0\mathbf{i} + 0\mathbf{j}$ ◄—

$\rightarrow (2 + a + b) = 0$ and ◄— *Adding the **i**-components and **j**-components.*

$\qquad (3a + b + 4) = 0$ *Equating coefficients of **i** and **j**.*

i.e. $\quad a + b = -2$ *Subtract the equations to eliminate b.*

$\qquad 3a + b = -4$ ◄—

$\qquad \rightarrow 2a = -2 \rightarrow a = -1 \rightarrow b = -1.$

(2)

$\qquad$ ◄——— *Draw a vector triangle.*

$P^2 + Q^2 = (2\sqrt{5})^2 = 20$ ◄——— *By Pythagoras.*

Draw another vector triangle, with the arrows 'following each other'; note that the angle is 120° not 60°.

$P^2 + Q^2 - 2PQ\cos120° = \left(2\sqrt{7}\right)^2 = 28$ ◄—— *By cosine rule.*

$P^2 + Q^2 + PQ = 28$ ◄——— *Cos120° = –0.5.*

So, $P^2 + Q^2 = 20$ and

$\qquad P^2 + Q^2 + PQ = 28$

$\rightarrow PQ = 8$ ◄——— *Subtracting the equations.*

So, $P^2 + Q^2 + 2PQ = 36$ ◄——— *Adding – this gives a neat solution as 36 is a perfect square.*

$\rightarrow (P + Q)^2 = 36$

$\rightarrow P + Q = 6$ i.e. $Q = 6 - P$ ◄—— *P + Q = –6 is impossible as P and Q are magnitudes.*

So, $P(6 - P) = 8$ ◄——

$\rightarrow P^2 - 6P + 8 = 0$ *Substituting for Q in PQ = 8.*

$\rightarrow (P - 2)(P - 4) = 0 \rightarrow P = 2$ or 4

$\qquad\qquad\qquad\qquad Q = 4$ or 2. ◄——— *There are two symmetrical solutions possible.*

(3) (a)

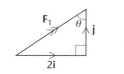

$\qquad$ ◄——— *A simple diagram is all that is needed.*

$\tan\theta = 2 \rightarrow \theta = 63.4°.$

(b) $\quad \mathbf{R} = (2\mathbf{i} + \mathbf{j}) + (a\mathbf{i} + 2a\mathbf{j}) = (2 + a)\mathbf{i} + (1 + 2a)\mathbf{j}$

So, $(1 + 2a) = 0 \quad$ i.e. $a = -0.5$ ◄——— ***R** parallel to **i** → **j**-component of **R** is zero.*

So, $\mathbf{R} = 1.5\mathbf{i}$

Hence, $|\mathbf{R}| = 1.5$

The magnitude of **R** is 1.5 N.

Answers

3.2 Kinematics

(1) $(\uparrow)\ 0 = 5u - 4.9.5^2 \rightarrow u = 24.5$
$(\uparrow)\ 0 = 24.5^2 - 2gh \rightarrow h = 30.625$
Maximum height is 30.625 m.

Choose a positive direction, here upwards.
Using $s = ut + \frac{1}{2}at^2$.
Using $v^2 = u^2 + 2as$.
Do not include units in your working but give the final answer with units.

(2) (a) $s = \frac{1}{2}(14 + 34)20 = 480$
The car travels 480 m.

Using $s = \frac{1}{2}(u + v)t$.

(b) $34 = 14 + 20a \rightarrow a = 1$
$240 = 14t + \frac{1}{2}t^2$
$t^2 + 28t - 480 = 0$
$(t + 40)(t - 12) = 0$
$t = -40$ or $t = 12$
The car takes 12 s.

First we need to find a, using $v = u + at$.
Using $s = ut + \frac{1}{2}at^2$.
Multiplying through by 2 to clear fractions and collecting terms.
Include both solutions and then reject where appropriate.

(3) Suppose the first stone hits the ground after t sec.
$h = \frac{1}{2}gt^2$
$h = 19.6(t - 1) + \frac{1}{2}g(t - 1)^2$
$\frac{1}{2}gt^2 = 19.6(t - 1) + \frac{1}{2}g(t - 1)^2$
$t^2 = 4(t - 1) + (t - 1)^2$
$2t - 1 = 4t - 4$
$2t = 3 \rightarrow t = 1.5 \rightarrow h = 11.025$
The height of the tower is 11.025 m.

Using $s = ut + \frac{1}{2}at^2$ for the first stone.
Using $s = ut + \frac{1}{2}at^2$ for the second stone.
Although we want h, it is easier to find t and then find h.
Dividing through by $\frac{1}{2}g$.
Multiplying out and cancelling terms.

(4) (a)

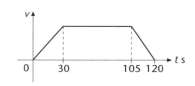

(b) Distance $= 2145 \rightarrow$ area $= 2145$
$\rightarrow \frac{1}{2}(120 + 75)v = 2145$
$\rightarrow 195v = 4290$
$\rightarrow v = 22$
So acceleration is $22/30 = 0.73$ m s^{-2}.

Using the Trapezium Rule.

(5) (a) $s_A = t + \frac{1}{2}2t^2 = t + t^2$

$s_B = 16(t-4) + \frac{1}{2}(t-4)^2$

We need $s_B > s_A$

i.e. $16(t-4) + \frac{1}{2}(t-4)^2 > t + t^2$

$32t - 128 + t^2 - 8t + 16 > 2t + 2t^2$

$0 > t^2 - 22t + 112$

$0 > (t-8)(t-14)$

$8 < t < 14$

i.e. B overtakes A after 8 s and A overtakes B after 14 s.

(b) When $t = 8$, $s_A = 8 + 8^2 = 72$

Cars are 72 m from O.

Using the result above.

(c) When $t = 14$, $s_B = 14 + 14^2 = 210$

Cars are 210 m from O.

Using the result above.

(6) (a) and **(b)**

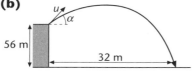

56 m, 32 m

A simple diagram helps to clarify the situation.

$(\rightarrow)$ $32 = 4u\cos\alpha$ → $u\cos\alpha = 8$

$(\uparrow) -56 = 4u\sin\alpha - \frac{1}{2} \times 9.8 \times 4^2$

→ $u\sin\alpha = 5.6$

So $\tan\alpha = \frac{5.6}{8}$ → $\alpha = 35.0°$

and $u = 9.77$.

Using $s = ut$ ($a = 0$ horizontally).

Using $s = ut + \frac{1}{2}at^2$ with upwards positive.

Dividing the two equations to eliminate u.

(c) $(\uparrow)$ $v = u\sin\alpha - 9.8 \times 4$

$= 5.6 - 39.2 = -33.6$

So speed $= \sqrt{8^2 + 33.6^2}\,\text{m s}^{-1}$

$= 34.5$ m s^{-1}.

First find the vertical component of the velocity when the stone hits the sea, using $v = u + at$; the negative sign indicates it is going downwards.
The horizontal component is 8 from above.

(d)

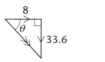

A simple diagram helps to clarify the situation.

$\tan\theta = 33.6/8$ → $\theta = 76.6°$ (3 s.f.)

The stone hits the sea at 76.6° to the horizontal.

Answers

(7) (a)

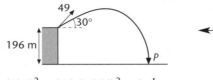

$(\uparrow)\ 0^2 = (49\sin 30°)^2 - 2gh$

$\rightarrow h = 30.625$

$\rightarrow$ height above sea $= 30.625 + 196 = 226.625$ m.

(b) $-196 = 49\sin 30° t - \tfrac{1}{2}gt^2$

$t^2 - 5t - 40 = 0$

$t = \tfrac{1}{2}\left(5 \pm \sqrt{185}\right)$

i.e. $t = 9.3$ or -4.3

Ball hits sea after 9.3 s.

(c) $s = 49\cos 30° \times 9.3 = 395$

P is 395 m from the foot of the cliff.

(8) (a) When $t = 0$, $v = \frac{45}{2} = 22.5$ km h^{-1}.

(b) $0 = \frac{15}{2}(3 + 4t - 4t^2)$

i.e. $4t^2 - 4t - 3 = 0$

$(2t + 1)(2t - 3) = 0$

$t = -0.5$ or 1.5

The train leaves A at 11.59.5 am and arrives at B at 12.01.5 pm.

(c) $a = \mathrm{d}v/\mathrm{d}t = \frac{15}{2}(4 - 8t) = 0$ at maximum velocity

$\rightarrow t = 0.5 \rightarrow v_{max} = \frac{15}{2}(3 + 2 - 1) = 30$ km h^{-1}.

(d) $s = \int \frac{15}{2}(3 + 4t - 4t^2)\,\mathrm{d}t = \frac{15}{2}(3t + 2t^2 - \frac{4}{3}t^3) + c.$

When $t = -0.5$, $s = 0 \rightarrow c = \frac{25}{4}$

When $t = 1.5$, $s = 40\frac{\mathrm{km}}{60} = \frac{2}{3}$ km

Average velocity $= \frac{2}{3} \times 30$ km h^{-1} $= 20$ km h^{-1}.

3.3 Statics

(1) (a)

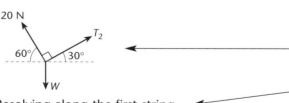

Resolving along the first string,

$20 = W\cos 30° \rightarrow W = 23.1$ N (3 s.f.).

(b) Resolving along the second string,

$T_2 = W\cos 60°$

$\rightarrow T_2 = 23.094 \times 0.5 = 11.5$ N (3 s.f.).

Examiner's tips
A simple diagram is a good idea.
Applying $v^2 = u^2 + 2as$ vertically, upwards positive.
This is the height above the point of projection.
Note $s = -198$; s is displacement not distance travelled.
Dividing through by -4.9 and collecting terms.
It doesn't factorise so use the formula.
Using $s = ut$ ($a = 0$ horizontally).
The train stops at both so put $v = 0$.
We need to find the distance AB.
Note that, because the time units are different, s is in $\frac{\mathrm{km}}{60}$.
A clear force diagram is essential.
Exploiting the fact that the two strings are perpendicular – if this were not the case then we would normally resolve vertically and horizontally and then have to solve two simultaneous equations.
To obtain 3 s.f. answers the working should be to at least 4 s.f.

(2)

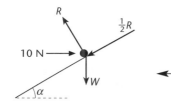

Examiner's tips

Since the particle is about to slip up the plane, the friction is limiting and is acting down the plane.

$R(\rightarrow)$, $10 - \frac{1}{2}R\cos\alpha = R\sin\alpha$

i.e. $10 - 0.4R = 0.6R$

$R = 10$ N

Resolving horizontally.

$\sin\alpha = 0.6$ and $\cos\alpha = 0.8$.

$R(\uparrow)$, $-\frac{1}{2}R\sin\alpha + R\cos\alpha = W$

$-5 \times 0.6 + 10 \times 0.8 = W$

5 N$= W$.

Resolving vertically.

(3)

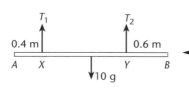

A simple diagram showing all the forces is essential.

$R(\uparrow)$, $T_1 + T_2 = 10g = 98$

$M(X)$, $1.4\,T_2 = 10g \times 0.8$

$T_2 = 56 \rightarrow T_1 = 42$

The tensions are 56 N and 42 N.

Since the rod is in equilibrium we can resolve in any direction.

Taking moments about a point through which an unknown force passes is usually a good idea.

(4) Let the centre of mass of the rod be x m from Q.

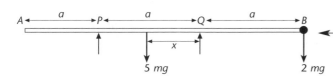

A simple diagram is essential.

$M(Q)$, $5\,mg\,x = 2\,mg\,a$

$x = \frac{2a}{5}$

Hence the distance of the centre of mass of the rod from B is $\frac{7a}{5}$.

When the rod is about to tilt about Q the reaction at P will be zero since the contact at P will be about to disappear.

Answers

3.4 Dynamics

(1) (a)

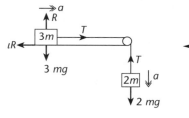

Examiner's tips

A clear force diagram, showing all the forces and any accelerations are absolutely crucial to the solution.

For B,

$1 = \frac{1}{2}a2^2 \rightarrow a = 0.5$.

Using $s = ut + \frac{1}{2}at^2$.

(b) For A,

$R = 3\,mg$

$T - \mu R = 3ma$

i.e. $T - \mu 3\,mg = 3ma$ ......**[1]**

We must consider each mass separately; friction is limiting as there is motion.

For B,

$2\,mg - T = 2ma$ ......**[2]**

Note that we have resolved in the direction of the acceleration for each mass.

[1] + [2],

$2\,mg - \mu 3\,mg = 5ma$

i.e. $g(2 - 3\mu) = 5a$ ......**[3]**

From part **(a)**

$9.8(2 - 3\mu) = 2.5$

$\rightarrow \qquad \mu = 0.58$.

(c) B hits the floor with speed $v = 0.5 \times 2 = 1$

Using $v = u + at$.

For A, $-\mu 3\,mg = 3ma \rightarrow a = -\mu g$

A common error is to put 'ma' on the RHS.

$0 = 1^2 - 2\mu g\,s \rightarrow s = 0.088$.

Since this is less than 0.2 m A does not reach the pulley.

(2) (a)

$\xrightarrow{u} \qquad \xleftarrow{3u}$

$P\left(2m\right) \qquad \left(km\right)Q \qquad \longrightarrow +ve$

$\xleftarrow{\frac{u}{2}} \qquad \xrightarrow{v}$

A clear diagram showing all the information is essential.

$2mu - km3u = -2m\left(\frac{u}{2}\right) + kmv$

Using conservation of momentum.

$3mu - 3kmu = kmv$

$3u(1 - k) = v$

$v > 0 \rightarrow (1 - k) > 0 \rightarrow k < 1$.

Using the fact that the direction of motion of Q is reversed (i.e. $v > 0$).

(b)

$I = 2\,m\frac{u}{2} - (-2mu)$

$= 3mu.$

Draw a diagram showing the velocities and the impulse.

Note that the impulse–momentum equation is a vector equation and attention must be paid to signs.

(3) (a)

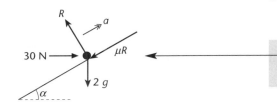

Draw a clear force diagram showing the forces and the acceleration.

$R\ (\nwarrow)$, $R - 30\sin\alpha - 2g\cos\alpha = 0$

$\to R = 18 + 15.68 = 33.68$

Normal reaction is 33.7 N (3 s.f.).

Resolving perpendicular to the acceleration.

(b) $R\ (\nearrow)$, $30\cos\alpha - 0.2R - 2g\sin\alpha = 2a$

$24 - 6.736 - 11.76 = 2a$

$a = 2.75$ (3 s.f.).

Resolving parallel to the acceleration.

Questions with model answers

For help see Revise AS Study Guide pages 94 and 95

C grade candidate – mark scored 4/7

(1)   A consumer group is investigating the use of the telephone within a particular town. The number of telephone units, t, used in a particular three month period by a random sample of 250 households were collated and summarised in a group frequency table. In order to simplify the arithmetic the data in the table were coded such that $x = \frac{1}{10}(t - 290)$, giving

$\Sigma fx = -40$ and $\Sigma fx^2 = 1075$.

(a) Find estimates of the mean and the variance of the number of telephone units used in that three month period in the town.   **[5]**

$\bar{x} = \frac{\Sigma fx}{250} = -0.16$ *and* $\bar{t} = 10\bar{x} + 290$

$\rightarrow t = -1.6 + 290 = 288.4$

$\sigma_x^2 = \frac{1075}{250} - (-0.16)2 = 2.0675.$

(b) Suggest two ways in which the accuracy of your estimates could be improved.   **[2]**

Examiner's Commentary

Correct answer for $\bar{x}$ and has rearranged the coding formula.

Correct – check the order of magnitude makes sense.

*Correct but candidate now forgets to multiply by 100 to find σ_t^2, **4/5 scored.***

*The candidate is unable to answer – possible ways would be to take a larger sample and/or use the actual data rather than group it, **0/2 scored.***

A grade candidate – mark scored 15/17

(2)   The average weekly incomes, in £, of households in 11 regions of the UK are given below:

254, 251, 268, 297, 359, 289, 266, 261, 247, 259, 219

(a) Find the median and the lower and upper quartiles.   **[3]**

219, 247, 251, 254, 259, 261, 266, 268, 289, 297, 359

$Q_2 = 261;\ Q_1 = 251;\ Q_3 = 289$

(b) On graph paper, draw a box plot to represent these data.   **[3]**

200 210 220 230 240 250 260 270 280 290 300 310 320 330 340 350 360

(c) Identify a possible outlier.   **[1]**

A possible outlier is 359.

Correct – put the data in numerical order.

*All correct, **3/3 scored.***

*All correct, **3/3 scored.** There must be a scale.*

*Correct, **1/1 scored.***

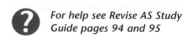

For help see Revise AS Study Guide pages 94 and 95

A grade candidate continued

(d) Find the mean and standard deviation of these data.　　**[6]**

x	x^2
219	47961
247	61009
251	63001
254	64516
259	67081
261	68121
266	70756
268	71824
289	83521
297	88209
359	128881
2970	814880

$$x = \frac{2970}{11}$$
$$= 270$$

$$sd = \sqrt{\left(\frac{814880}{11} - 270^2\right)}$$
$$= 34.35 \ (2 \ d.p.)$$

Examiner's Commentary

*All correct; the candidate has shown all the working which is a good idea – check your answer using the stats functions on your calculator, **6/6 scored**.*

(e) Further investigation suggests that the £359 value should in fact be £309. Without further calculation say what effect this change would have on

(i) the standard deviation

(ii) the interquartile range,

explaining your answers.　　**[4]**

(i) The standard deviation will decrease since the spread of the data will decrease.

Correct; the standard deviation takes all values into account.

(ii) The interquartile range will decrease for the same reason.

*Incorrect; the IQR is a measure of the spread of the 'middle half' of the data which is unaffected by the change; in general the IQR is not changed by extreme values, **2/4 scored**.*

Exam practice questions

Answers on pp. 53–57

4.1 Representing data

(1) The stem and leaf diagram below summarises the data giving the number of minutes, to the nearest minute, that a random sample of 65 trains from Guildford were late arriving at a main line station.

Minutes late	0	2 means 2	Totals
0	2 3 3 3 4 4 4 4 5 5 5 5 5 5	(14)
0	6 6 6 7 7 8 8 8 9	(9)
1	0 0 0 2 2 3 3 4 4 4 5	( )
1	6 6 7 7 8 8 8 9 9	( )
2	1 2 2 3 3 3 3 4	( )
2	6	( )
3	3 4 4 5	(4)
3	6 8	(2)
4	1 3	(2)
4	7 7 9	(3)
5	2 4	(2)

(a) Write down the missing values. **[2]**

(b) Find the median and the quartiles of these times. **[3]**

(c) Find the 65th percentile. **[2]**

(d) On graph paper construct a box plot for these data, showing the scale clearly. **[3]**

A random sample of trains arriving at the same main line station from Reading had a minimum value of 15 minutes late and a maximum of 30 minutes. The quartiles were 18, 22 and 27 minutes.

(e) On the same graph paper, using the same scale, construct a box plot for these data. **[3]**

(f) Compare and contrast the journeys from Guildford and Reading based on these data. **[3]**

4.2 Probability

(1) (a) Show that, for any two events A and B:
$P(A \cup B) = P(A) + P(B) - P(A \cap B)$. **[2]**

(b) Express in words the meaning of $P(A \mid B)$. **[1]**

Given that A and B are independent events,

(c) express $P(A \cap B)$ in terms of $P(A)$ and $P(B)$, [1]

(d) show that A' and B are also independent. [3]

In a school, 60 pupils are studying one or more of the three subjects: Biology, Chemistry and Physics. Of these, 25 are studying Biology, 26 are studying Chemistry, 44 are studying Physics, 10 are studying Biology and Chemistry, 15 are studying Chemistry and Physics and 16 are studying Biology and Physics.

(e) Write down the probability that a student, chosen at random from those studying Physics, is also studying Chemistry. [1]

(f) Determine whether or not the events 'studying Biology' and 'studying Chemistry' are independent. [2]

A student is chosen at random from all 60 students.

(g) Find the probability that the chosen student is studying all three subjects. [5]

(2) A golfer has six different clubs in his golf bag, only one of which is correct for the shot about to be played. The probability that the golfer plays a good shot if the correct club is chosen is $\frac{3}{5}$ and the probability of a good shot if the incorrect club is chosen is $\frac{1}{4}$. The golfer chooses a club at random and plays the shot.

(a) Find the probability that a good shot is made. [4]

(b) The golfer plays a good shot. What is the probability that he chose the correct club? [3]

(c) Find the probability that an incorrect club was used given that a bad shot was made. [3]

(d) Comment on the model that the golfer chooses a club at random. [1]

(3) The events A and B are independent and $P(A \mid B) = \frac{3}{4}$ and $P(B) = P(B' \cap A')$. By letting $P(B) = x$ and forming an equation in x, find

(a) $P(B)$, [5]

(b) $P(B' \cap A)$. [2]

(c) Write down $P(B \mid A)$. [1]

(4) A bag contains 3 red, 4 white and 5 blue balls. Three balls are selected at random from the bag, without replacement. Find the probability that the three balls are of different colours.

Exam practice questions

4.3 Discrete random variables

(1) A random number generator in a certain computer game produces values which can be modelled by the discrete random variable R whose probability function is given by

$$P(R = r) = kr! \quad r = 0, 1, 2, 3, 4$$

where k is a constant.

(a) Show that $k = \frac{1}{34}$. [2]

(b) Sketch the probability distribution of R. [2]

(c) Find $E(R)$ and Var(R). [4]

Two independent values of R, R_1 and R_2, are generated.

(d) Find $P(R_1 = R_2)$. [3]

(e) Given that $R_1 = R_2$, find the probability that $R_1 = R_2 = 4$. [3]

(2) The random variable X has the following distribution:

x:	1	2	3
$P(X = x)$:	p	q	p

(a) Find $E(X)$. [5]

Given that Var$(X) = 0.75$,

(b) find the values of p and q. [1]

(3) A spinner is made from the disc in the diagram and the random variable N represents the number that it falls on after being spun.

(a) Find the distribution of N. [2]

(b) Write down $E(N)$. [1]

(c) Find Var(N). [3]

Sophie and Tom use the spinner to play a board game. Sophie's score is given by the random variable $2N - 1$ and Tom's score is given by the random variable $3N - 3$.

(d) Show that the mean score for each player is the same. [3]

(e) Find the variance of Sophie's score. [2]

4.4 The Normal distribution

(1) The random variable X is distributed Normally with mean 5 and variance 4. Calculate

    (a) $P(X = 64)$ **[1]**

    (b) $P(X > 0)$ **[3]**

    (c) $P(|X - 5| > 3)$.   **[5]**

(2) The random variable Y is Normally distributed with mean μ and variance σ^2. Given that $P(Y > 58.37) = 0.02$ and $P(Y < 40.85) = 0.01$, calculate μ and σ^2.   **[10]**

(3) A machine produces gaskets for engines. The engine manufacturer's specification is that the thickness of the gaskets should be between 5.45 mm and 5.55 mm, and the diameter should lie between 8.45 mm and 8.54 mm. The machine produces gaskets whose thicknesses are Normally distributed with mean 5.5 mm and variance 0.0004 mm^2 and whose diameters are independently Normally distributed with mean 8.54 mm and variance 0.0025 mm^2.
Find, to 1 decimal place, the percentage of gaskets produced which will not meet:

    (a)   the thickness specifications,   **[6]**

    (b)   the diameter specifications.   **[6]**

4.5 Correlation and regression

(1) In order to assess the potential of ten prospective new pupils, a selective school sets a Mathematics test and a Verbal Reasoning (VR) test. Their results were as follows, where x represents their Mathematics score and y represents their VR score:

x	4	21	12	11	15	13	29	17	15	15
y	30	39	22	25	28	37	45	20	32	34

    (a)   Calculate the value of the product–moment correlation coefficient, r, between x and y, given that $\Sigma x^2 = 2696$, $\Sigma y^2 = 10288$ and $\Sigma xy = 5014$.   **[6]**

A concerned parent of one of the prospective new pupils suggests that, in future, perhaps the school should set only one test in order to reduce the pressure on the children hoping to gain entry to the school. In response to this suggestion the Headteacher states that the school would like to retain both tests since they measure different abilities.

    (b)   Comment on the Headteacher's statement in the light of your answer to part **(a)**.   **[2]**

Exam practice questions

(2) Two branches of a small retail chain of grocery shops were situated close to each other in a small town. It was thought that sales in one may possibly be affecting the sales in the other and since no other shops nearby sold fresh vegetables it was decided to compare sales of fresh vegetables to see if this was true. Sales of vegetables were recorded weekly in pounds for seven weeks; the data collected are recorded in the following table:

Week	1	2	3	4	5	6	7
Shop A	380	402	370	365	410	392	385
Shop B	560	543	564	573	550	554	540

(a) Using a method of coding, calculate the product–moment correlation. **[9]**

(b) Comment on your result. **[2]**

(3) A Head of Mathematics needs to make predictions about the final A level grade each of his students will achieve. To do this he decides to look at the marks obtained in mock examinations. In Mathematics he believes there is a linear relationship between the mark obtained in a mock examination and the final mark obtained. To investigate this he looks at the results from last year; the mock mark and final mark are given in the table below:

Mock mark x	18	26	28	34	36	42	48	52	54	60
Final mark y	54	64	54	62	68	70	76	66	76	74

(a) Draw a scatter diagram to illustrate these data. **[2]**

(b) Calculate the regression line in the form $y = a + bx$. **[9]**

(c) What final marks might be expected to be gained by students obtaining 16, 30, 50, 85 in their mock examination? **[3]**

(d) Comment on the validity of these predictions. **[2]**

Answers

4.1 Representing data

(1) (a)   11, 9, 8, 1.

(b) $\frac{1}{4}(65) = 16.25$ so Q_1 is 17th fig. i.e. 6 ◄────── *If $\frac{1}{4}n$ is not an integer, round up.*

$\frac{1}{2}(65) = 32.5$ so Q_2 is 33rd fig. i.e. 14 ◄────── *If $\frac{1}{2}n$ is not an integer, round up.*

$\frac{3}{4}(65) = 48.75$ so Q_3 is 49th fig. i.e. 23. ◄────── *If $\frac{3}{4}n$ is not an integer, round up.*

(c) $\frac{65}{100}(65) = 42.25$ so P_{65} is 43rd fig. i.e. 19.

(d) $Q_1 - 1.5(Q_3 - Q_1) = 6 - 1.5 \times 17 = -19.5$ ◄────── *So no outliers at lower end.*

$Q_3 + 1.5(Q_3 - Q_1) = 23 + 1.5 \times 17 = 48.5$. ◄────── *So 49, 52, 54 are all outliers.*

◄────── *LH whisker starts at lowest value.*

◄────── *RH whisker stops at 48.5 with the three outliers marked individually.*

(e) $Q_1 - 1.5(Q_3 - Q_1) = 18 - 1.5 \times 9 = 4.5$ ◄────── *So no outliers at lower end.*

$Q_3 + 1.5(Q_3 - Q_1) = 27 + 1.5 \times 9 = 40.5$. ◄────── *So no outliers at upper end.*

(f)   Both likely to be late but the lateness for the Guildford service is more inconsistent and could be quite severe. The spread for the Reading service is smaller and hence it may be easier to plan for.

4.2 Probability

(1) (a) $n(A \cup B) = n(A) + n(B) - n(A \cap B)$

Dividing through by $n(\varepsilon)$ gives the result. ◄────── *Since $A \cap B$ is counted twice.*

(b)   Probability of A given B.

(c) $P(A \cap B) = P(A).P(B)$.

(d) $P(A').P(B) = (1 - P(A)).P(B)$ ◄────── *We must show that $P(A')P(B) = P(A' \cap B)$.*

$= P(B) - P(A).P(B)$

$= P(B) - P(A \cap B)$ ◄────── *Using the independence of A and B.*

$= P(A' \cap B)$ hence A' and B are independent.

(e) $\frac{15}{44}$.

(f) $P(B) = 25/60$; $P(B \mid C) = 10/26$ – as these are different B and C are not independent. ◄────── *This is a more intuitive definition of independence – as the outcome of B is affected by the outcome of C they are not independent.*

Answers

(g)

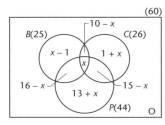

Examiner's tips

Using a Venn Diagram is the right approach – always work out from the centre when entering the data.

Suppose there are x pupils doing all three then $44 + 10 - x + x - 1 + x + 1 = 60$, so $x = 6$; hence probability is $\frac{6}{60} = \frac{1}{10}$.

(2) (a) $P(\text{correct club}) = \frac{1}{6}$ and $P(\text{incorrect club}) = \frac{5}{6}$ ⟵ *Since there are 6 clubs to choose from.*

$P(\text{good shot}) = \left(\frac{1}{6} \times \frac{3}{5}\right) + \left(\frac{5}{6} \times \frac{1}{4}\right) = \frac{37}{120}$. ⟵ *A tree diagram could have been used.*

(b) $P(\text{correct club given a good shot}) = \dfrac{\left(\frac{1}{6} \times \frac{3}{5}\right)}{\left(\frac{37}{120}\right)} = \frac{12}{37}$.

(c) $P(\text{bad shot}) = \left(\frac{1}{6} \times \frac{2}{5}\right) + \left(\frac{5}{6} \times \frac{3}{4}\right) = \frac{83}{120}$ ⟵ *Or we could have used $\left(1 - \frac{37}{120}\right)$.*

$P(\text{incorrect club given a bad shot}) = \dfrac{\left(\frac{5}{6} \times \frac{3}{4}\right)}{\left(\frac{83}{120}\right)}$ ⟵ *This is a conditional probability.*

$= \frac{75}{83}$.

(d) This is totally unrealistic.

(3) (a) $P(A) = P(A \mid B) = \frac{3}{4}$ since A and B are independent and A' and B' are also independent

so $P(B' \cap A') = P(B').P(A')$

So, $x = (1 - x). \frac{1}{4} \to x = \frac{1}{5}$ i.e. $P(B) = \frac{1}{5}$.

As A and B do not affect each other $P(A) = P(A \mid B)$.

(b) $P(B' \cap A) = P(B').P(A) = \frac{4}{5} \times \frac{3}{4} = \frac{3}{5}$. ⟵ *As B and A are independent.*

(c) $P(B \mid A) = P(B) = \frac{1}{5}$. ⟵ *As B and A are independent.*

(4) Prob $= \frac{3}{12} \times \frac{4}{11} \times \frac{5}{10} \times 3! = \frac{3}{11}$. ⟵ *There are 3! ways of choosing a red, white and blue ball. The denominators of the fractions decrease by 1 each time as the selection is without replacement.*

4.3 Discrete random variables

(1) (a) $k(0! + 1! + 2! + 3! + 4!) = 1$ ⟵ *The sum of the probabilities is 1.*

$k(1 + 1 + 2 + 6 + 24) = 1 \to k = \frac{1}{34}$.

(b)

r	0	1	2	3	4
$P(R = r)$	$\frac{1}{34}$	$\frac{1}{34}$	$\frac{2}{34}$	$\frac{6}{34}$	$\frac{24}{34}$

This is the probability distribution of R.

This is a histogram showing the distribution of R – note that the total area is 1.

(c) $E(R) = 0 \times \frac{1}{34} + 1 \times \frac{1}{34} + 2 \times \frac{2}{34} + 3 \times \frac{6}{34} + 4 \times \frac{24}{34}$

$= \frac{119}{34} = 3\frac{17}{34} = 3.5$

$\text{Var}(R) = E(R^2) - [E(R)]^2$

$= 0^2 \times \frac{1}{34} + 1^2 \times \frac{1}{34} + 2^2 \times \frac{2}{34} + 3^2 \times \frac{6}{34} + 4^2 \times \frac{24}{34} - (3.5)^2$

$= 0.897$ (3 s.f.).

Note that it is possible to find both $E(R)$ and $\text{Var}(R)$ on a calculator by treating the probabilities as frequencies – use $\bar{x}$ for $E(R)$ and $(\sigma_x)^2$ for $\text{Var}(R)$.

(d) $P(0, 0) + P(1, 1) + P(2, 2) + P(3, 3) + P(4, 4)$
$= \{P(0)\}^2 + \{P(1)\}^2 + \{P(2)\}^2 + \{P(3)\}^2 + \{P(4)\}^2$
$= 0.535$ (3 s.f.).

Since R_1 and R_2 are independent.

(e) $P(R_1 = R_2 = 4) = \left(\frac{24}{34}\right)^2 = 0.498$

A conditional probability is required.

$\frac{P(R_1 = R_2 = 4)}{P(R_1 = R_2)}$

$= \frac{0.49826...}{0.5346...} = 0.932$ (3 s.f.).

Note that we have to work to at least 4 s.f. to be sure of obtaining an answer which is correct to 3 s.f.

(2) (a) $E(X) = p + 2q + 3p = 4p + 2q$
Also $2p + q = 1$,
So $E(X) = 2(2p + q) = 2$.

Since the sum of the probabilities is 1.

(b) $\text{Var}(X) = E(X^2) - [E(X)]^2$
$0.75 = p + 4q + 9p - 2^2$
$\rightarrow 10p + 4q = \frac{19}{4}$
And $2p + q = 1$;
Solving gives $p = \frac{3}{8}$ and $q = \frac{1}{4}$.

(3) (a)

n:	1	2	3
$P(N = n)$:	$\frac{3}{8}$	$\frac{2}{8}$	$\frac{3}{8}$

The probabilities are proportional to the angles by symmetry.

(b) $E(N) = 2$. — *By symmetry.*

(c) $\text{Var}(N) = \frac{1}{8}(3 + 8 + 27) - 2^2 = 0.75$. — *$\text{Var}(N) = E(N^2) - [E(N)]^2$.*

(d) $E(2N - 1) = 2E(N) - 1 = 3$
$E(3N - 3) = 3E(N) - 3 = 3$. — *$E[aN + b] = aE[N] + b$.*

(e) $\text{Var}(2N - 1) = 4\text{Var}(N) = 3$. — *$\text{Var}[aN + b] = a^2\text{Var}[N]$.*

4.4 The Normal distribution

(1) (a) $P(X = 64) = 0$. — *Since there is no area – a surprising result.*

(b) $P(X > 0) = P\left(Z > \frac{(0 - 5)}{2}\right)$ — *Standardising the variable.*
$= P(Z < 2.5)$ — *By symmetry – draw a diagram.*
$= \Phi(2.5)$
$= 0.9938$.

(c) $P(|X - 5| > 3) = P(X > 8 \text{ or } X < 2)$
$= P(X > 8) + P(X < 2)$
$= 1 - P(X < 8) + P(X < 2)$
$= 1 - P\left(Z < \frac{(8-5)}{2}\right) + P\left(Z < \frac{(2-5)}{2}\right)$
$= 1 - P(Z < 1.5) + P(Z < -1.5)$
$= 1 - P(Z < 1.5) + 1 - P(Z < 1.5)$
$= 2 - 2\Phi(1.5)$
$= 0.1336$.

It is best to write it without the $||$ signs although here, as $\mu = 5$, X has to be more than 3 above or 3 below the mean, i.e. more than 1.5 SD above or below the mean, i.e. prob = $2(1 - \Phi(1.5))$.

Answers

(2)

$$P(Y > 58.37) = 0.02$$
$$1 - P(Y < 58.37) = 0.02$$
$$P(Y < 58.37) = 0.98$$
$$P\left(Z < \frac{(58.37 - \mu)}{\sigma}\right) = 0.98$$
$$\Phi\left(\frac{(58.37 - \mu)}{\sigma}\right) = 0.98$$
$$\left(\frac{(58.37 - \mu)}{\sigma}\right) = 2.05$$
$$\mu + 2.05\sigma = 58.37 \qquad (1)$$
$$P(Y < 40.85) = 0.01$$
$$P\left(Z < \frac{(40.85 - \mu)}{\sigma}\right) = 0.01$$
$$\Phi\left(\frac{(40.85 - \mu)}{\sigma}\right) = 0.01$$
$$\Phi\left\{-\left(\frac{(40.85 - \mu)}{\sigma}\right)\right\} = 0.99$$
$$\mu - 2.33\sigma = 40.85 \qquad (2)$$
$$4.38\sigma = 17.52$$
$$\sigma = 4 \text{ i.e. } \sigma^2 = 16$$
so
$$\mu - 9.32 = 40.85$$
$$\mu = 50.17.$$

- *From a diagram.*
- *Standardising the variable.*
- *Rearranging.*
- *Standardising the variable.*
- *We cannot read back from the tables as 0.01 < 0.5; using the symmetry of the graph.*
- *(1) – (2) to eliminate μ.*
- *Substitute for σ in (2).*

(3) (a)

$$P(5.45 < T < 5.55) = P\left(-\frac{0.05}{0.02} < Z < \frac{0.05}{0.02}\right)$$
$$= P(-2.5 < Z < 2.5) = \Phi(2.5) - \Phi(-2.5)$$
$$= \Phi(2.5) - \{1 - \Phi(2.5)\} = 2\Phi(2.5) - 1 = 0.9876$$

Hence, answer is $(1 - 0.9876) \times 100\% = 1.2\%$ (to 1 d.p.).

- *First find the probability that they do meet the specification.*
- *Using the symmetry of the graph.*

(b)

$$P(8.45 < D < 8.54) = P\left(-\frac{0.09}{0.05} < Z < 0\right)$$
$$= P(-01.8 < Z < 0) = \Phi(0) - \Phi(-1.8)$$
$$= 0.5 - 1 + \Phi(1.8) = 0.4641$$

Hence, answer is $(1 - 0.4641) \times 100\% = 53.6\%$ (to 1 d.p.).

- *First find the probability that they do meet the specification.*
- *Using the symmetry of the graph.*

4.5 Correlation and regression

(1) (a) $\Sigma x = 152 \quad \Sigma y = 312 \quad \Sigma x^2 = 2696 \quad \Sigma y^2 = 10288 \quad \Sigma xy = 5014.$

$$S_{xx} = 2696 - \frac{152^2}{10} = 385.6$$

$$S_{yy} = 10288 - \frac{312^2}{10} = 553.6$$

$$S_{xy} = 5014 - \frac{152 \times 312}{10} = 271.6$$

$$r = \frac{S_{xy}}{\sqrt{(S_{xx} S_{yy})}} = \frac{271.6}{\sqrt{(385.6 \times 553.6)}} = 0.588 \text{ (3 s.f.).}$$

- *It is best to set out the calculation like this.*
- *You can use a calculator to check this but you risk losing all the marks if you just put down the answer, and it's wrong.*

(b) This value of r suggests a high positive correlation between the two tests, i.e. one test should do.

(2) (a)   Let $a = A - 370$ $b = B - 540$

a	b	a^2	b^2	ab
10	20	100	400	200
32	3	1024	9	96
0	24	0	576	0
−5	33	25	1089	−165
40	10	1600	100	400
22	14	484	196	308
15	0	225	0	0
114	104	3458	2370	839

$S_{aa} = 3458 - \frac{114^2}{7} = 1601.43$

$S_{bb} = 2370 - \frac{104^2}{7} = 824.86$

$S_{ab} = 839 - \frac{114 \times 104}{7} = -854.71$

$r = \frac{S_{ab}}{\sqrt{(S_{aa}S_{bb})}} = \frac{-854.71}{\sqrt{(1601.43 \times 824.86)}} = -0.744$ (3 s.f.).

Examiner's tips

You can use a calculator to check this.

(b)   This value of r indicates a high negative correlation between the sales in each shop – i.e. if they're up in one they're likely to be down in the other and vice versa.

(3) (a)

(b) $\Sigma x = 398$ $\Sigma y = 664$ $\Sigma x^2 = 17524$ $\Sigma y^2 = 44680$ $\Sigma xy = 27268$

$S_{xx} = 17524 - \frac{398^2}{10} = 1683.6$

$S_{xy} = 27268 - \frac{398 \times 664}{10} = 840.8$

It is best to set out the calculation like this.

$b = \frac{840.8}{1683.6} = 0.4994 = 0.5$ (2 s.f.)

It is best to set out the calculation like this.

$a = \bar{y} - b\bar{x} = 66.4 - 0.5 \times 39.8 = 46.5$

i.e. $y = 46.5 + 0.5x$.

(c)   When $x = 30$, $y = 61.5$

When $x = 50$, $y = 71.5$

Both 30 and 50 are well within the range of the data collected; so these predictions are fairly safe.

When $x = 16$, $y = 54.5$

16 is just outside the range of the data so this value is fairly reliable.

However, $x = 85$ cannot be used, with any degree of confidence, to predict as it is well outside the range of the data.

Questions with model answers

For help see Revise AS Study
Guide pages 111 – 113

C grade candidate – mark scored 9/12

(1) The highest common factor (HCF) of two numbers A and B in which $A > B$ can be found using the following algorithm.

Step 1:   Input A and B.

Step 2:   Take B from A.

Step 3:   If $A > B$ then go back to step 2.

Step 4:   If $A = B$ then print A and stop.

Step 5:   Reverse A and B and return to step 2.

(a) Demonstrate this algorithm when $A = 189$ and $B = 153$.   **[8]**

A:	189	36	153	117	81	45	9	36	27	18	9
B:	153	153	36	36	36	36	36	9	9	9	9
Steps:	1	2,3,4	5	2,3	2,3	2,3	2,3,4	5	2,3	2,3	2,3,4

Examiner's Commentary

This is correct – candidates are strongly advised to write out each step as shown. This is not only so that the examiner can see the steps but so that the work can be checked. However, crucially, the printout is not given! **7/8 scored.**

(b) Demonstrate what happens if A and B are interchanged and explain why it is necessary to have the initial condition $A > B$. **[4]**

A:	153	−36	189	225
B:	189	189	−36	−36
Steps:		2	5	2

Because after step 2 A would be a negative number.

This is so, but the process would not fall down there, because step 3 is satisfied, step 4 is not and so step 5 would take place, resulting in A being positive and B negative; in this case $A = 189$ and $B = −36$. Now successive subtractions of B from A will only increase the value of A and an everlasting cycle will be initiated. The candidate should be aware that more is required for full marks, **2/4 scored.**

A grade candidate – mark scored 13/15

(2)   When two friends arrive at an airport there are two check-in desks for their flight. There is a separate queue for each desk.
It may be assumed that the distribution of check-in times at each desk is as follows:

Examiner's Commentary

Time taken (min)	1	2	3	4	5
Probability	0.1	0.4	0.2	0.2	0.1

(a)   If a set of random digits taken in pairs are to be used to simulate the check-in times in which 00–09 will represent a check-in time of 1 minute, then give a table to show how this may be done for the distribution above.   **[1]**

There are other ways to do the simulation which would be accepted, but candidates should be encouraged to devise the method that is simplest to explain and simplest to use, 1/1 scored.

00–09 for 1 min, 10–49 for 2 min, 50–69 for 3 min, 70–89 for 4 min, 90–99 for 5 min.

(b)   The first desk has two groups in the queue with the first just starting the check-in procedure. Use the random numbers given to simulate six times the time of check-in for the two groups and hence the waiting time for the friends for each simulation.

Simulation number	Random number for first group	Random number for second group
1	73	91
2	38	07
3	03	64
4	24	10
5	53	70
6	38	70

Calculate the mean waiting time from your six simulations.   **[5]**

Group 1 times: 4   2   1   2   3   2
Group 2 times: 5   1   3   2   4   4
Totals: 9   3   4   4   7   6
Mean waiting time = 5.5 min.

Good candidates should have no trouble with this computation in their heads, but all are strongly advised to write everything down. The missing number here is the sum of totals, but since the answer is correct this is unimportant, 5/5 scored.

Questions with model answers

A grade candidate continued

(c) The second queue had three groups waiting to check in. Instead of joining the first queue with two groups, the friends decide to join both queues, one in each queue. The first to be served will then check both of them in.

(i) Use the random numbers given to simulate six times the progress of the longer queue. **[3]**

Simulation number	Random number for first group	Random number for second group	Random number for third group
1	21	26	67
2	32	73	85
3	86	28	74
4	60	77	95
5	35	41	04
6	32	93	45

Group 1 times: 2   2   4   3   2   2
Group 2 times: 2   4   2   4   2   5
Group 3 times: 3   4   4   5   1   2
Totals: 7   10   10   12   5   9

(ii) Combine the results with those from the simulation from **(b)** to obtain six values for the friends' waiting time if they use this strategy.

Give the mean of the six waiting times. **[3]**

Minimum time: 7   3   4   4   5   6   *gives mean 4.83 mins.*

(d) Give three ways in which this simulation experiment could be improved or made more realistic. **[3]**

'Groups' could be anything from one to four people and so group times for check-in will be different.
Set up another simulation process to account for the group size (say 1 to 4).
Do it again with different random numbers.

Exam practice questions

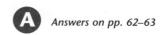

Answers on pp. 62–63

(1) A retailer in children's toys decides to stock two types of sets of building bricks. The cheaper costs him £5 and comes in a box measuring $20 \times 30 \times 40$ cm. The more expensive costs him £7 and comes in a box measuring $30 \times 40 \times 50$ cm. He decides to allocate £500 to the purchase, 3 m^3 of space in the shop to storage, and to sell them for £7 and £10 respectively.

    (a) Write down the objective function, given that the retailer buys x cheap and y expensive boxes. **[2]**

    (b) Write down two constraints. **[2]**

    (c) Solve the problem given that the retailer wishes to maximise his profits. **[4]**

    (d) What would be the effect of trying to increase the number of boxes in the order by 10% while retaining the constraints? **[4]**

(2) The table shows activities involved in a building project, with their duration (in hours) and immediate predecessors.

Activity	A	B	C	D	E	F	G	H
Immediate predecessors	–	–	B	A, C	A, B	E	D, E	G
Duration	5	6	2	5	6	7	6	3

    (a) Complete the activity network. **[4]**

    (b) Find the early and late times by performing a forward and backward pass. **[4]**

    (c) Give the critical path and the minimum time to completion. **[2]**

(3) The diagram represents the roads joining eight villages, labelled A–H. The numbers give distances in km.

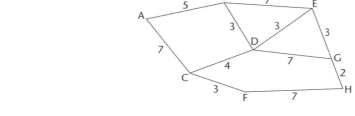

    (a) Use Dijkstra's algorithm to find a shortest route from A to H. Explain the method carefully and show all your working. Give the shortest route and its length. **[5]**

    (b) Find the time that John takes to travel by the quickest route if he drives on average 40 kilometres per hour. **[2]**

    (c) One particular day John finds that the road from B to D is flooded and there is a fallen tree on the road from C to D. He knows that each of these obstructions adds 5 minutes to the journey times. Find the shortest time that John will take on the day with these obstructions and give the route that he should take. **[3]**

(4) A school list consists of 897 names, all given in alphabetical order. Demonstrate how the binary search algorithm can be used to locate the name LING if it actually appears as number 743 in the table, giving the positions of the names which will be checked and the number of tests required. **[5]**

Answers

(1) (a) The profit on each box is £2 and £3 respectively, so $P = 2x + 3y$.

Think what it is that is to be maximised. In this case it is the profit, not just income. So $P = (7 - 5)x + (10 - 7)y$.

(b) For space:
$$0.024x + 0.060y \leqslant 3 \rightarrow 2x + 5y \leqslant 250$$
For cost:
$$5x + 7y \leqslant 500.$$

The units are mixed (and reasonably so!) and you should take care to be consistent. Either use cubic metres throughout by converting the dimensions of the box, or use the dimensions in centimetres and convert the 3 m³ to cm³.

(c) The two constraint lines meet at (68.2, 22.7) but the solution must be an integer point. Investigation of points near will give (69, 22) as the value that maximises the profit of £204.

The two inequations should be rewritten as equations and solved. Sometimes this will give the point that maximises the Objective Function, but in many cases (including this one!) the values of x and y need to be integers. All points around the intersection of the lines should be checked. In this case (68, 22) is OK but there may be a better value. (69, 22) is OK (and is the best solution. (68, 23) is outside at least one of the constraints. Don't forget that it is the profit that is required, not just the point that gives the maximum profit!

(d) The number of boxes is 69 + 22 = 91. An increase of 10% in the number of boxes would mean $x + y = 100$. The only point in the feasible region is (100, 0). This would give a storage space of 2.4 m³ being required and an outlay of £500. If all were sold then the profit would be £200.

The number of boxes is simply $x + y$; 10% more than 91 is 100, giving $x + y = 100$. Ensure that the constraints are met and find the profit.

(2) (a) and **(b)**

Complete the diagram carefully with earliest and latest times.

(c) Critical path B, C, D, G, H. Minimum time: 22 hours.

Check the line where the difference between latest time of one and the earliest time of the one before equals the length of the activity.

(3) (a)

The shortest distance is via ABDEGH and is 16 km.

The diagram shows Dijkstra's algorithm applied using one notation system. You may have learnt another system, but the result will be the same! Fill in the boxes carefully, checking each step to ensure that you have it right.

(b) The time taken is 24 minutes.

Change each length to time, or take the total distance of 16 km at 40 km per hour, giving 24 minutes.

(c) The route will now be round either
perimeter and will take 25.5 minutes.

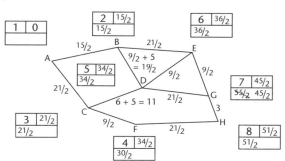

(4) The binary search will locate the middle name, i.e. $\frac{(897 + 1)}{2} = 449$.
This name will be lower in the list than LING.
The next name to be identified will be $\frac{(449 + 897)}{2} = 673$.
This name will be lower in the list than LING.
The next name to be identified will be $\frac{(673 + 897)}{2} = 785$.
This name will be higher in the list than LING.
The next name to be identified will be $\frac{(673 + 785)}{2} = 729$.
This name will be lower in the list than LING.
The next name to be identified will be $\frac{(729 + 785)}{2} = 757$.
This name will be higher in the list than LING.
The next name to be identified will be $\frac{(757 + 729)}{2} = 743$.
The name is found after 6 iterations.

Questions with model answers

 For help see Revise A2 Study Guide pages 22–24 and 18–21

C grade candidate – mark scored 3/6

1. Find the coordinates of the centre and radius of the circle with equation **[6]**

$$2x^2 + 2y^2 - 8x + 12y - 5 = 0.$$

$x^2 + y^2 - 4x + 6y - 5 = 0$
$x^2 - 4x + 4 + y^2 + 6y + 9 = 5 + 4 + 9$
$(x - 2)^2 + (y + 3)^2 = 18$
so, centre is $(2, 3)$ and radius is $\sqrt{18}$.

Examiner's Commentary

Correct – we need the coefficients of x^2 and y^2 to both be 1 but the candidate has forgotten to divide the 5 by 2 also. Completing the square is the correct method.

Incorrect; the centre should be $(2, -3)$ and the radius is wrong due to the earlier error, 3/6 scored.

If time permits it is always a good idea to multiply out the final expression and ensure that it is the same or equivalent to the original equation given in the question.

A grade candidate – mark scored 10/12

(2) Given that

$$f(x) \equiv \frac{11 - 5x^2}{(x + 1)(2 - x)}$$

(a) find constants A and B such that $f(x) \equiv \dfrac{A}{(x + 1)} + \dfrac{B}{(2 - x)}$ **[5]**

$$\frac{11 - 5x^2}{(x+1)(2-x)} = 5 + \frac{A}{(x+1)} + \frac{B}{(2-x)}$$

$11 - 5x^2 \quad = 5(x+1)(2-x) + A(2-x) + B(x+1)$
$x = 2: \qquad -9 = 3B \to B = -3$
$x = -1: \qquad 6 = 3A \to A = 2.$

Given that x is so small that x^3 and higher powers of x may be ignored,

(b) find the series expansion of $f(x)$ in ascending powers of x up to and including the term in x^2. **[7]**

$f(x) = 5 + 2(1 + x)^{-1} - 3(2 - x)^{-1}$
$\quad = 5 + 2(1 + x)^{-1} - 3.2\left(1 - \frac{x}{2}\right)^{-1}$
$\quad = 5 + 2(1 + (-1)x + (-1)(-2)x^2)$
$\qquad - 6\left(1 + (-1)\left(-\frac{x}{2}\right) + (-1)(-2)\left(-\frac{x}{2}\right)^2\right)$
$\quad = 5 + 2(1 + x + x^2) - 6\left(1 + -\frac{x}{2} + \frac{x^2}{2}\right)$
$\quad = 1 - x - x^2.$

Correct – clearing the fractions is the best way to start. This is an identity so we can put in any value of x – the two values chosen make each bracket in turn zero, 5/5 scored.

Correct – writing the partial fractions in this form allows the candidate to use the Binomial Theorem – note that the 2 must be taken out as a factor first but the candidate has forgotten that it should be 2^{-1}. Writing it out in full to start with is always a good idea, particularly the brackets around the negative and fractional terms, 5/7 scored.

Incorrect final answer due to the earlier error – the candidate could have checked the constant in the final answer by putting $x = 0$ i.e. $f(0) = \frac{11}{2}$ but putting $x = 0$ in the series expansion gives 1!

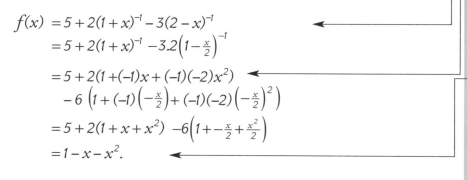

Exam practice questions

 Answers on pp. 69–75

6.1 Algebra

(1) The function f is given by
$$f(x) = px^3 + 11x^2 + 2px - 5.$$
When $f(x)$ is divided by $(x + 2)$, the remainder is 15.

    (a)   Show that $p = 2$. **[3]**

    (b)   Factorise $f(x)$ completely. **[3]**

    (c)   Hence find the solutions of the equation
$$f(x) = (x + 5)(x + 1).$$ **[4]**

(2) Given that the polynomial $P(x)$ is divisible by $(x - a)^2$, show that $P'(x)$ is divisible by $(x - a)$. **[4]**
The polynomial $x^4 + x^3 - 12x^2 + px + q$ is divisible by $(x + 2)^2$.
Find the values of p and q. **[7]**

(3) The polynomial $P(x)$ is given by $P(x) = 6x^3 + ax^2 + x - 2$, where a is a constant. Given that $(x + 2)$ is a factor of $P(x)$,

    (a)   find the value of a. **[3]**

Using this value of a,

    (b)   express $1/P(x)$ in partial fractions **[9]**

    (c)   find the value of $\frac{d^2}{dx^2}(1/P(x))$ at $x = 0$. **[10]**

6.2 Coordinate geometry

(1) Given that the line $y = mx$ is a tangent to the circle with equation $x^2 + y^2 - 6x - 6y + 17 = 0$, find the possible values of m. **[8]**

(2) **(a)**   Sketch the graph of $y = (x + 2)(x - 1)$. **[3]**

    (b)   Hence sketch the graph of $y = 1/(x + 2)(x - 1)$, giving the equations of any asymptotes. **[4]**

(3) Find the equation of the normals to the curve whose equation is $y^2 + 3xy + 4x^2 = 37$ at the points where $x = 4$. **[15]**

(4)
$$f(x) = ax^3 + bx^2 + cx + d.$$
The curve with equation $y = f(x)$ has gradient 4 at the point with coordinates $(0, -5)$.

    (a)   Find the values of c and d. **[4]**

The remainder when $f(x)$ is divided by $(x + 2)$ is 15 and the remainder when $f(x)$ is divided by $(x - 1)$ is 12.

    (b)   Find the values of a and b. **[6]**

Exam practice questions

(5) A curve C has parametric equations
$$x = 2t + 3, \; y = t^3 - 4t.$$
The point P has parameter $t = -1$ and the line l is the tangent to C at P.
The line l also cuts the curve at the point Q.

(a) Show that an equation for l is $x + 2y - 7 = 0$. [9]

(b) Find the coordinates of Q. [6]

6.3 Series

(1) **(a)** $f(x) \equiv \dfrac{1+2x}{(6x^2 + 1)(1 - 3x)}$

Express $f(x)$ in partial fractions. [6]

(b) Hence, or otherwise, expand $f(x)$ in ascending powers of x as far as the term in x^3. [5]

(c) State the range of values of x for which the expansion is valid. [3]

(2) **(a)** Expand $(4 - 2x)^{1/2}$ in ascending powers of x, up to the term in x^3. [4]

(b) State the range of values of x for which your expansion is valid. [2]

(c) By putting $x = \frac{3}{8}$ in your expansion, find $\sqrt{13}$ to 1 d.p. [4]

6.4 Differentiation

(1) Given that $2xy = e^x + e^{2y}$, find $\frac{dy}{dx}$. [5]

(2) The number of bacteria, N, present in a sample, t hours after noon on a particular day, is given by
$$N = 2000e^{kt}, \text{ where } k \text{ is a constant}$$
When $t = 0$, the rate of increase of the number of bacteria is 200 per hour.

(a) Find the value of k. [4]

(b) Find the number of bacteria present at 6 pm on the same day. [2]

(3) **(a)** Differentiate with respect to x

(i) $\ln(2x^2)$ [2]

(ii) $x^2\cos 4x$ [2]

(iii) $4\sin^3(2x)$. [2]

(b) Find the gradient of the curve with equation $x^2y + y^2 = 6$ at the point (1, 2). [6]

(4) The curve, C, has equation $y = \frac{2x}{1 + x^2}$.

    (a)   Show that $\frac{dy}{dx} = \frac{2(1 - x^2)}{(1 + x^2)^2}$.   **[3]**

    (b)   Hence find the coordinates of the stationary points and distinguish between them.   **[6]**

    (c)   Sketch the curve.   **[3]**

6.5 Integration

(1)

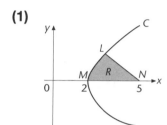

The curve C shown in the diagram has parametric equations
$$x = t + \frac{1}{t},\, y = t - \frac{1}{t},\, t > 0$$

    (a)   Find $\frac{dy}{dx}$ in terms of t.   **[4]**

    (b)   Show that the Cartesian equation of C is $x^2 - y^2 = 4$.   **[3]**

The point L on C has coordinates $\left(2\frac{1}{2},\, 1\frac{1}{2}\right)$ and the point M on C has coordinates $(2, 0)$. The point N has coordinates $(5, 0)$. The region R is bounded by the lines MN and LN and the arc LM of the curve C. The region R is rotated through 2π about the x-axis to form a solid of revolution.

    (c)   Find the volume of the solid, leaving your answer in terms of π.   **[6]**

(2)

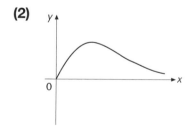

The diagram shows the curve with equation $y = 2xe^{-x/2}$

    (a)   Find the coordinates of the turning point on the curve.   **[6]**

The region R is bounded by the curve, the x-axis and the line $x = 2$. Find

    (b)   the area of R.   **[6]**

Exam practice questions

(3) **(a)** Use the substitution $u^2 = x - 1$ to find $\int x\sqrt{x-1}\ dx$. **[8]**

(b) Use integration by parts to find $\int x(2x+1)^{-1/2}dx$. **[4]**

(c) Find $\int \tan^2 2x\ dx$. **[3]**

(4) **(a)** Express $\dfrac{1}{(x+3)(1+x)}$ in partial fractions. **[5]**

(b) Find the solution of the differential equation
$$(x+3)(1+x)\frac{dy}{dx} = y,\ x > -1$$
given that when $x = 1$, $y = 2$. **[7]**

6.6 Vectors

(1) A line L_1 passes through the point P with position vector 5**i** + 3**j** and the point Q with position vector −2**i** − 4**j** + 7**k**.

(a) Find an equation of the line. **[3]**

A second line L_2 has equation **r** = **i** − 3**j** − 4**k** + λ (**i** + 2**j** + 3**k**) where λ is a parameter.

(b) Show that L_1 and L_2 are perpendicular. **[2]**

(c) Show that L_1 and L_2 meet and find the position vector of the point of intersection. **[5]**

The point R has position vector 2**i** − **j** − **k**.

(d) Show that R lies on L_2. **[1]**

The point S is the image of R after reflection in the line L_1.

(e) Find the position vector of S. **[3]**

(2) The position vectors of the points A and B are 2**i** − 3**j** + 3**k** and 5**i** + **j** + c**k** respectively, where c is a constant. The point C is such that $OABC$ is a rectangle where O is the origin.

(a) Find the value of c. **[5]**

(b) Write down the position vector of C. **[1]**

(c) Find, in Cartesian form, an equation of the line BC. **[4]**

(3) The line L_1 is parallel to the vector (2**i** − **j**) and passes through the point with position vector (**i** + **j** − **k**). The line L_2 passes through the points with position vectors (**i** + **j** − **k**) and (3**i** − 2**j** + **k**). Find the acute angle between L_1 and L_2. **[7]**

Answers

6.1 Algebra

(1) (a)
$$f(-2) = 15$$ — *Using the Remainder Theorem.*
$$-8p + 44 - 4p - 5 = 15$$ — *Putting $x = -2$.*
$$12p = 24$$
$$p = 2.$$

(b)
$$f(x) = 2x^3 + 11x^2 + 4x - 5$$ — *Now look for a value of x for which $f(x) = 0$.*
$$f(1) = 2 + 11 + 4 - 5 \neq 0$$ — *Try $x = 1, -1, 2, -2$ etc.*
$$f(-1) = -2 + 11 - 4 - 5 = 0 \rightarrow (x + 1)$$ — *By the Factor Theorem.*
is a factor of $f(x)$
$$\rightarrow f(x) = (x + 1)(2x^2 + 9x - 5)$$ — *By inspection – the first and third terms are easy!*
$$= (x + 1)(x + 5)(2x - 1).$$

(c)
$$(x + 1)(x + 5)(2x - 1) = (x + 1)(x + 5)$$ — *Don't multiply out!*
$$(x + 1)(x + 5)(2x - 1) - (x + 1)(x + 5) = 0$$ — *Collecting the terms.*
$$(x + 1)(x + 5)\{(2x - 1) - 1\} = 0$$ — *Taking out the common factor.*
$$(x + 1)(x + 5)(2x - 2) = 0$$
$$x = -1 \text{ or } -5 \text{ or } 1.$$

(2)
Let $P(x) = (x - a)^2 Q(x)$ — *Since $P(x)$ is divisible by $(x - a)^2$.*
$$P'(x) = 2(x - a)Q(x) + (x - a)^2 Q'(x)$$ — *Using the Product Rule.*
$$= (x - a)\{2Q(x) + (x - a)Q'(x)\}$$ — *Taking out the common factor.*
Hence $P'(x)$ is divisible by $(x - a)$.

Let $f(x) = x^4 + x^3 - 12x^2 + px + q$
Then $f'(x) = 4x^3 + 3x^2 - 24x + p$
is divisible by $(x + 2)$ — *Using the above result.*
So $f'(-2) = 0$ i.e. $-32 + 12 + 48 + p = 0$ — *Using the Factor Theorem.*
i.e. $p = -28$
Also $f(-2) = 0$ i.e. $16 - 8 - 48 + 56 + q = 0$ — *Using the Factor Theorem.*
i.e. $q = -16$.

(3) (a)
$$P(-2) = 0$$ — *Using the Factor Theorem.*
$$-48 + 4a - 2 - 2 = 0$$
$$a = 13.$$

(b)
$$P(x) = (x + 2)(6x^2 + x - 1)$$ — *First we need to factorise $P(x)$.*
$$= (x + 2)(3x - 1)(2x + 1)$$
$$\frac{1}{P(x)} = \frac{1}{(x + 2)(3x - 1)(2x + 1)}$$
$$= \frac{A}{(x + 2)} + \frac{B}{(3x - 1)} + \frac{C}{(2x + 1)}$$ — *These are the appropriate partial fractions.*
$$1 = A(3x - 1)(2x + 1) + B(x + 2)(2x + 1) + C(x + 2)(3x - 1)$$ — *Multiplying through by $(x + 2)(3x - 1)(2x + 1)$.*
$$x = \tfrac{1}{3}: 1 = B\tfrac{7}{3}.\tfrac{5}{3} \rightarrow B = \tfrac{9}{35}$$ — *Choose values of x to make each factor zero.*
$$x = -\tfrac{1}{2}: 1 = C\tfrac{3}{2}.-\tfrac{5}{2} \rightarrow C = -\tfrac{4}{15}$$
$$x = -2 : 1 = A. -7. -3 \rightarrow A = \tfrac{1}{21}$$
$$\frac{1}{P(x)} = \frac{1}{105}\left(\frac{5}{(x + 2)} + \frac{27}{(3x - 1)} - \frac{28}{(2x + 1)}\right).$$

Answers

(c)

$$\frac{1}{P(x)} = \frac{1}{105}\left(5(x+2)^{-1} + 27(3x-1)^{-1} - 28(2x+1)^{-1}\right)$$

Express fractions as powers before differentiating.

$$\frac{d}{dx}\left(\frac{1}{P(x)}\right) = \frac{1}{105}\left(-5(x+2)^{-2} - 81(3x-1)^{-2} + 56(2x+1)^{-2}\right)$$

Differentiating, using the chain rule.

$$\frac{d^2}{dx^2}\left(\frac{1}{P(x)}\right) = \frac{1}{105}\left(10(x+2)^{-3} + 486(3x-1)^{-3} - 224(2x+1)^{-3}\right)$$

Differentiating again.

$$\left(\frac{1}{P(0)}\right)'' = \frac{1}{105}\left(10 \times 2^{-3} + 486(-1)^{-3} - 224(1)^{-3}\right) = -\frac{27}{4}$$

Putting $x = 0$.

when $x = 0$.

6.2 Coordinate geometry

Solving the two equations simultaneously.

(1)

$$x^2 + (mx)^2 - 6x - 6(mx) + 17 = 0$$
$$(1 + m^2)x^2 - 6(1 + m)x + 17 = 0$$

If $y = mx$ is a tangent this equation will have repeated roots so '$b^2 = 4ac$'.

$$\rightarrow 36(1 + m)^2 = 68(1 + m^2)$$
$$\rightarrow 4m^2 - 9m + 4 = 0$$
$$\rightarrow m = \frac{9 \pm \sqrt{17}}{8}.$$

Collecting terms; this equation does not factorise so use the quadratic formula.

(2) (a)

When $y = 0$

$$(x + 2)(x - 1) = 0 \rightarrow x = -2 \text{ or } 1$$

When $x = 0$, $y = -2$

There is a turning point at $x = \frac{-2+1}{2} = -\frac{1}{2}$

then $y = -\frac{9}{4}$

The curve is a parabola; find where it crosses the axes by putting $y = 0$ and $x = 0$.

The turning point is on the line of symmetry.

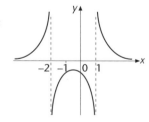

(b)

There are vertical asymptotes at $x = -2$ and $x = 1$ and a horizontal asymptote at $y = 0$.
The turning point is again on the line of symmetry
i.e. $x = -\frac{1}{2}$ but $y = -\frac{4}{9}$ (the reciprocal of the value in **(a)**).

Where $y = 0$, since as $x \rightarrow \infty$, $y \rightarrow 0$.

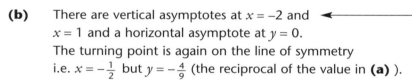

Note the relationship between the two graphs. They cross when $y = \frac{1}{y}$ i.e. when $y = \pm 1$.

(3) $y^2 + 3xy + 4x^2 = 37$

When $x = 4$, $y^2 + 12y + 64 = 37$

$y^2 + 12y + 27 = 0$

$(y + 3)(y + 9) = 0$

$y = -3$ or -9 i.e. points are $A(4, -3)$
and $B(4, -9)$

First find all the points where $x = 4$.

$y^2 + 3xy + 4x^2 = 37$

$2y\frac{dy}{dx} + 3\left(x\frac{dy}{dx} + y\right) + 8x = 0$

$\frac{dy}{dx}(2y + 3x) = -(8x + 3y)$

$\frac{dy}{dx} = -\frac{(8x + 3y)}{(2y + 3x)}$

At A, $\frac{dy}{dx} = -\frac{23}{6}$; at B, $\frac{dy}{dx} = \frac{5}{6}$

At A, gradient of normal is $\frac{6}{23}$

So $y - -3 = \frac{6}{23}(x - 4)$

$23y - 6x + 93 = 0$

At B, gradient of normal is $-\frac{6}{5}$

So $y - -9 = -\frac{6}{5}(x - 4)$

$5y + 6x + 21 = 0$.

Differentiating with respect to x.

Collecting terms in $\frac{dy}{dx}$ on one side and the rest on the other.

Using the coordinates of A and B.

Invert the gradient of the tangent and change the sign.

Using $y - y_1 = m(x - x_1)$.

Check that the coordinates of A satisfy this equation.

Check that the coordinates of B satisfy this equation.

(4) (a) $f(0) = -5 \rightarrow d = -5$

$f'(x) = 3ax^2 + 2bx + c$

$f'(0) = 4 \rightarrow c = 4$.

First find the derivative of $f(x)$.

(b) $f(-2) = 15$

$-8a + 4b - 8 - 5 = 15$ i.e. $-2a + b = 7$

$f(1) = 12$

$a + b + 4 - 5 = 12$ i.e. $a + b = 13$

$3a = 6 \rightarrow a = 2$

$\rightarrow b = 11$.

Using the Remainder Theorem.

Using the Remainder Theorem.

Subtracting the equations to eliminate b.

(5) (a) When $t = -1$, $x = 1$ and $y = 3$

$\frac{dy}{dt} = 3t^2 - 4$; $\frac{dx}{dt} = 2$

When $t = -1$, $\frac{dy}{dt} = -1$

So $\frac{dy}{dx} = -\frac{1}{2}$

Equation of tangent is $y - 3 = -\frac{1}{2}(x - 1)$

$2y - 6 = -x + 1$ i.e. $x + 2y - 7 = 0$.

First find the coordinates of P.

We need to find the gradient of the curve at P.

$\frac{dy}{dx} = \frac{dy}{dt} / \frac{dx}{dt}$.

Using $y - y_1 = m(x - x_1)$.

(b) $(2t + 3) + 2(t^3 - 4t) - 7 = 0$

$2t^3 - 6t - 4 = 0$ i.e. $t^3 - 3t - 2 = 0$

$(t + 1)^2(t - 2) = 0$

So $t = 2$ is the parameter of Q

So $x = 7$, $y = 0$

i.e. Q has coordinates $(7, 0)$.

Substituting for x and y in the tangent equation; two of the roots of this equation are -1 (twice) since the tangent touches the curve at $t = -1$.

The third factor can be found by inspection.

Answers

6.3 Series

(1) (a) Let $\dfrac{1+2x}{(6x^2+1)(1-3x)} \equiv \dfrac{Ax+B}{(6x^2+1)} + \dfrac{C}{(1-3x)}$

$1 + 2x = (Ax+B)(1-3x) + C(6x^2+1)$ ←

$x = \frac{1}{3}: \ \frac{5}{3} = C.\frac{5}{3} \rightarrow C = 1$

$-3A + 6C = 0 \rightarrow A = 2$ ←

$1 = B + C \rightarrow B = 0$ ←

So $f(x) = \dfrac{2x}{(6x^2+1)} + \dfrac{1}{(1-3x)}$.

(b) $f(x) = 2x(1 + 6x^2)^{-1} + (1 - 3x)^{-1}$ ←

$= 2x(1 - 6x^2 + \ldots) +$

$\left(1 + (-1)(-3x) + \dfrac{(-1)(-2)(-3x)^2}{2!} + \dfrac{(-1)(-2)(-3)(-3x)^3}{3!} + \ldots\right)$ ←

$= 2x - 12x^3 + 1 + 3x + 9x^2 + 27x^3 + \ldots$

$= 1 + 5x + 9x^2 + 15x^3 + \ldots$

(c) First series valid for $|6x^2| < 1$ i.e. $|x| < \frac{1}{\sqrt{6}}$ ←

Second series valid for $|-3x| < 1$ i.e. $|x| < \frac{1}{3}$

So both valid for $|x| < \frac{1}{3}$ as $\frac{1}{3} < \frac{1}{\sqrt{6}}$ (as $3 > \sqrt{6}$).

(2) (a) $(4 - 2x)^{1/2} = \{4(1 - \frac{1}{2}x)\}^{1/2}$ ←

$= 4^{1/2}(1 - \frac{1}{2}x)^{1/2}$

$= 2\left[1 + \left(\frac{1}{2}\right)\left(\frac{-1}{2}x\right) + \dfrac{\left(\frac{1}{2}\right)\left(\frac{-1}{2}\right)\left(\frac{-1}{2}x\right)^2}{2!} + \dfrac{\left(\frac{1}{2}\right)\left(\frac{-1}{2}\right)\left(\frac{-3}{2}\right)\left(\frac{-1}{2}x\right)^3}{3!} + \ldots\right]$ ←

$= 2\left(1 - \frac{1}{4}x - \frac{1}{12}x^2 - \frac{1}{128}x^3 + \ldots\right)$

$= 2 - \frac{1}{2}x - \frac{1}{6}x^2 - \frac{1}{64}x^3 + \ldots$

(b) $|-\frac{1}{2}x| < 1 \rightarrow |x| < 2$.

(c) $\left(4 - \frac{3}{4}\right)^{1/2} = 2 - \frac{3}{16} - \frac{9}{384} - \frac{27}{32768}$ ←

i.e. $\sqrt{13} = 2\left(2 - \frac{3}{16} - \frac{9}{384} - \frac{27}{32768}\right) = 3.576\ldots = 3.6$ (1 d.p.).

6.4 Differentiation

(1) $2\left(x\frac{dy}{dx} + y\right) = e^x + 2\frac{dy}{dx}e^{2y}$ ←

$\frac{dy}{dx}(2x - 2e^{2y}) = (e^x - 2y)$ ←

$\frac{dy}{dx} = (e^x - 2y)/(2x - 2e^{2y})$.

(2) (a) $\frac{dN}{dt} = 2000ke^{kt}$ ←

So, $200 = 2000k \rightarrow k = 0.1$.

(b) $N = 2000e^{0.6} = 3644$.

(3) (a) (i) $4x\left(\dfrac{1}{2x^2}\right) = \dfrac{2}{x}$ ← Using the chain rule; alternatively write $\ln(2x^2)$ as $2\ln x + \ln 2$, then differentiate.

(ii) $2x\cos 4x - 4x^2\sin 4x$ ← Using the product rule.

(iii) $24\sin^2 2x\cos 2x.$ ← Using the chain rule.

(b) $x^2\dfrac{dy}{dx} + 2xy + 2y\dfrac{dy}{dx} = 0$ ← Using the product and chain rules.

$\dfrac{dy}{dx}(x^2 + 2y) = -2xy$

$\dfrac{dy}{dx} = -2xy/(x^2 + 2y)$

When $x = 1$, $y = 2$, $\dfrac{dy}{dx} = -\dfrac{4}{5}$.

(4) (a) $\dfrac{dy}{dx} = \dfrac{(1+x^2).2 - 2x.2x}{(1+x^2)^2} = \dfrac{2(1-x^2)}{(1+x^2)^2}.$ ← Using the quotient rule.

(b) When $\dfrac{dy}{dx} = 0$, $(1-x^2) = 0$ ← Stationary points are points where the gradient is 0.

$\rightarrow x = \pm 1 \rightarrow y = \pm 1$

Points are (1, 1) and (−1, −1)

For x just less than 1, $\dfrac{dy}{dx} > 0$ ← To distinguish between the points consider the sign of $\dfrac{dy}{dx}$ either side of each point.

and for x just more than 1, $\dfrac{dy}{dx} < 0$

Hence (1, 1) is a maximum point.

For x just less than −1, $\dfrac{dy}{dx} < 0$

and for x just more than −1, $\dfrac{dy}{dx} > 0$

Hence (−1, −1) is a minimum point.

(c) When $x = 0$, $y = 0$

As $x \rightarrow +\infty$, $y \rightarrow 0^+$; as $x \rightarrow -\infty$, $y \rightarrow 0^-$

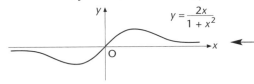

$y = \dfrac{2x}{1+x^2}$

Note that $f(x) = \dfrac{2x}{1+x^2}$ is an odd function since $f(-x) = -f(x)$. Hence the graph of $y = f(x)$ has rotational symmetry, order 2, about 0.

6.5 Integration

(1) (a) $\dfrac{dx}{dt} = 1 - \dfrac{1}{t^2}$; $\dfrac{dy}{dt} = 1 + \dfrac{1}{t^2}$

$\dfrac{dy}{dx} = \left(1 + \dfrac{1}{t^2}\right)\Big/\left(1 - \dfrac{1}{t^2}\right) = (t^2 + 1)/(t^2 - 1).$ ← $\dfrac{dy}{dx} = \dfrac{dy}{dt}\Big/\dfrac{dx}{dt}.$

(b) $x^2 = \left(t + \dfrac{1}{t}\right)^2 = t^2 + 2 + \dfrac{1}{t^2}$ ← To find the Cartesian equation we must eliminate the parameter t.

$y^2 = \left(t + \dfrac{1}{t}\right)^2 = t^2 - 2 + \dfrac{1}{t^2}$

Subtracting gives $x^2 - y^2 = 4$.

(c) $V = \pi\displaystyle\int_2^{5/2} y^2\,dx + \text{volume of cone}$

$= \pi\displaystyle\int_2^{5/2} (x^2 - 4)\,dx + \dfrac{1}{3}\pi\left(\dfrac{3}{2}\right)^2\left(\dfrac{5}{2}\right)$ ← Volume of a cone is $\dfrac{1}{3}\pi r^2 h.$

$= \pi\left[\dfrac{1}{3}x^3 - 4x\right]_2^{5/2} + \dfrac{15\pi}{8}$

$= \dfrac{29\pi}{12}.$

73

Answers

(2) (a) $y = 2xe^{-x/2}$

$\dfrac{dy}{dx} = 2e^{-x/2} - xe^{-x/2}$ ← *Using the product rule.*

$= (2 - x)\, e^{-x/2}$ ← *Factorising.*

At a turning point, $0 = (2 - x)e^{-x/2}$

$\rightarrow x = 2$, since $e^{-x/2} > 0$ for all x

When $x = 2$, $y = 4e^{-1}$ i.e. $(2, 4e^{-1})$ is the turning point.

(b) Area $= \displaystyle\int_0^2 y\, dx = \int_0^2 2xe^{-x/2}\, dx$

$= \left[2x.(-2e^{-x/2})\right]_0^2 - \displaystyle\int_0^2 2(-2e^{-x/2})\, dx$ ← *Using integration by parts.*

$= -8e^{-1} - \left[8e^{-x/2}\right]_0^2$

$= 8 - 16e^{-1}.$

(3) (a) $u^2 = x - 1 \rightarrow 2u\dfrac{du}{dx} = 1 \rightarrow 2u\,du = dx$ ← *First differentiate the substitution.*

$\displaystyle\int x\sqrt{(x-1)}\, dx = \int (u^2 + 1)\, u\, .2u\,du = 2\int u^4 + u^2\, du$ ← *Every part of the integral must be put in terms of u.*

$= 2\left(\tfrac{1}{5}u^5 + \tfrac{1}{3}u^3\right) + c$

$= \tfrac{2}{15}u^3\,(3u^2 + 5) + c$ ← *Factorise before changing back to x.*

$= \tfrac{2}{15}(x - 1)^{3/2}\,(3x - 3 + 5) + c$

$= \tfrac{2}{15}(x - 1)^{3/2}\,(3x + 2) + c.$ ← *The final answer must be in terms of x.*

(b) $I = x\,(2x + 1)^{1/2} - \displaystyle\int (2x + 1)^{1/2}\, dx$ ← *Using integration by parts – differentiating the x.*

$= x\,(2x + 1)^{1/2} - \tfrac{1}{3}(2x + 1)^{3/2} + c.$

(c) $\displaystyle\int \tan^2 2x\,dx = \int \sec^2 2x - 1\, dx = \tfrac{1}{2}\tan 2x - x + c.$ ← *Using $1 + \tan^2 x = \sec^2 x$.*

(4) (a) Let $\dfrac{1}{(x+3)(1+x)} = \dfrac{A}{(x+3)} + \dfrac{B}{(1+x)}$

$1 = A\,(1 + x) + B\,(x + 3)$ ← *Multiplying through by $(x + 3)(1 + x)$.*

$x = -1: 1 = 2B \rightarrow B = \tfrac{1}{2}$ ← *Choose x-values to make each factor 0.*

$x = -3: 1 = -2A \rightarrow A = -\tfrac{1}{2}$

$\dfrac{1}{(x+3)(1+x)} = \dfrac{-1/2}{(x+3)} + \dfrac{1/2}{(1+x)} = \dfrac{-1}{2(x+3)} + \dfrac{1}{2(1+x)}.$

(b) $\dfrac{dy}{y} = \dfrac{dx}{(x+3)(1+x)}$ ← *Separating the variables.*

$\displaystyle\int \dfrac{dy}{y} = \int \dfrac{-1}{2(x+3)} + \dfrac{1}{2(1+x)}\, dx$ ← *Using the partial fractions from part (a).*

$\ln y = \tfrac{1}{2}\{-\ln(x + 3) + \ln(1 + x)\} + \ln C$ ← *Add on $\ln C$.*

$2\ln y = \ln y^2 = \ln C'\dfrac{(1 + x)}{(x + 3)}$ ← *Where $\ln C' = 2\ln C$.*

$y^2 = C\dfrac{(1 + x)}{(x + 3)};$ ← *Combining and removing the logs.*

When $x = 1$, $y = 2$

$\rightarrow 4 = \dfrac{C}{2} \rightarrow C = 8$

$y^2 = \dfrac{8(1 + x)}{(x + 3)}.$

6.6 Vectors

(1) (a) Direction vector $= (5\mathbf{i} + 3\mathbf{j}) - (-2\mathbf{i} - 4\mathbf{j} + 7\mathbf{k}) = 7\mathbf{i} + 7\mathbf{j} - 7\mathbf{k}$

$\mathbf{r}_1 = (5\mathbf{i} + 3\mathbf{j}) + \mu(\mathbf{i} + \mathbf{j} - \mathbf{k})$. ◄

We can omit the 7s from the direction vector.

(b) $(\mathbf{i} + \mathbf{j} - \mathbf{k}).(\mathbf{i} + 2\mathbf{j} + 3\mathbf{k}) = 1 + 2 - 3 = 0$ ◄

Hence perpendicular.

Taking the scalar product of the direction vectors of the two lines.

(c) At intersection,

$(5\mathbf{i} + 3\mathbf{j}) + \mu(\mathbf{i} + \mathbf{j} - \mathbf{k}) = (\mathbf{i} - 3\mathbf{j} - 4\mathbf{k}) + \lambda(\mathbf{i} + 2\mathbf{j} + 3\mathbf{k})$ ◄

Equating the position vectors.

$\left.\begin{array}{l} 5 + \mu = 1 + \lambda \\ 3 + \mu = -3 + 2\lambda \\ 0 - \mu = -4 + 3\lambda \end{array}\right\}$ ◄

Equating coefficients of $\mathbf{i}$, $\mathbf{j}$ and $\mathbf{k}$.

$3 = -7 + 5\lambda$ ◄

$\rightarrow \lambda = 2 \rightarrow \mu = -2$

Adding the 2nd and 3rd equations.

Check that these values satisfy *all three* equations

p.v. of intersection point is $3\mathbf{i} + \mathbf{j} + 2\mathbf{k}$.

(d) When $\lambda = 1$ in the equation of L_2, $\mathbf{r} = 2\mathbf{i} - \mathbf{j} - \mathbf{k}$,

hence R lies on L_2.

(e)

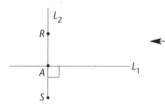

A simple diagram here is very helpful.

$\overrightarrow{AS} = \overrightarrow{RA}$

If A is the point of intersection of L_1 and L_2

then $RA = \mathbf{i} + 2\mathbf{j} + 3\mathbf{k}$, then p.v. of S is given by

$(3\mathbf{i} + \mathbf{j} + 2\mathbf{k}) + (\mathbf{i} + 2\mathbf{j} + 3\mathbf{k}) = 4\mathbf{i} + 3\mathbf{j} + 5\mathbf{k}$.

(2) (a) $\overrightarrow{OA}.\overrightarrow{AB} = 0$ ◄

$(2\mathbf{i} - 3\mathbf{j} + 3\mathbf{k}).\{(5\mathbf{i} + \mathbf{j} + c\mathbf{k}) - (2\mathbf{i} - 3\mathbf{j} + 3\mathbf{k})\} = 0$

$(2\mathbf{i} - 3\mathbf{j} + 3\mathbf{k}).(3\mathbf{i} + 4\mathbf{j} + (c - 3)\mathbf{k}) = 0$

$6 - 12 + 3c - 9 = 0$

$c = 5$.

Since $OABC$ is a rectangle, $\overrightarrow{OA}$ is perpendicular to $\overrightarrow{AB}$; hence their dot product is zero.

(b)

Draw a simple diagram; as $OABC$ is a rectangle, $\overrightarrow{OC} = \overrightarrow{AB}$.

$\overrightarrow{OC} = \overrightarrow{AB} = (3\mathbf{i} + 4\mathbf{j} + 2\mathbf{k})$.

(c) Direction vector of line $= OA$. ◄

Since $\overrightarrow{OA}$ is parallel to $\overrightarrow{BC}$.

Vector equation is $\mathbf{r} = (5\mathbf{i} + \mathbf{j} + c\mathbf{k}) + \lambda(3\mathbf{i} + 4\mathbf{j} + 2\mathbf{k})$

So $\frac{(x-5)}{3} = \frac{(y-1)}{4} = \frac{(z-5)}{2}$

is a possible Cartesian equation.

(3) Direction vector of L_2 is $(3\mathbf{i} - 2\mathbf{j} + \mathbf{k}) - (\mathbf{i} + \mathbf{j} - \mathbf{k})$

i.e. $(2\mathbf{i} - 3\mathbf{j} + 2\mathbf{k})$. ◄

$(2\mathbf{i} - 3\mathbf{j} + 2\mathbf{k}).(2\mathbf{i} - \mathbf{j}) = 7$

$\sqrt{(2^2 + 3^2 + 2^2)} \sqrt{(2^2 + 1^2)} \cos\theta = 7$

i.e. $\cos\theta = \frac{7}{\sqrt{85}}$

$\theta = 40.6°$.

The angle between the lines is the same as the angle between the direction vectors. Calculate the scalar product of the two direction vectors in two different ways and equate them.

Questions with model answers

For help see Revise A2 Study Guide pages 82 and 83

C grade candidate – mark scored 6/10

(1) A particle P moves in such a way that at time t seconds its position vector, **r** metres, relative to a fixed origin O, is given by

$$\mathbf{r} = ct^2\,\mathbf{i} + (t^3 - 4t)\mathbf{j}$$

where c is a positive constant. When $t = 2$, the speed of P is 10 m s^{-1}.

(a) Find the value of c. **[7]**

$$\mathbf{v} = 2ct\mathbf{i} + (3t^2 - 4)\mathbf{j}$$
When $t = 2$, $\mathbf{v} = 4c\mathbf{i} + 8\mathbf{j}$
$$10 = 16c^2 + 64$$
$$-54 = 16c^2$$

(b) Find the acceleration of P when $t = 2$. **[3]**

$$a = 2c\mathbf{i} + 6t^2\mathbf{j}$$
When $t = 2$,
$$a = 2c\mathbf{i} + 24\mathbf{j}$$

Examiner's Commentary

Correct start.
Also correct.
The candidate has forgotten to square the 10 and should realise, given that c^2 has come out to be negative, **4/7 scored.**

Another error – it should be 6t**j**.
Incorrect, but the method is correct and will receive credit, **2/3 scored.**

A grade candidate – mark scored 14/18

For help see Revise A2 Study Guide pages 82 and 83

Examiner's Commentary

(2) A bird leaves its nest for a short horizontal flight along a straight line, and then returns. In an initial model of the situation, the distance s metres, from the nest at time t seconds is given by

$$s = 25t - \frac{5}{2}t^2, \qquad 0 \leqslant t \leqslant 10.$$

(a) Find the value of s when $t = 2$. **[1]**

$$s = 25 \times 2 - \frac{5}{2} \times 2^2 = 40$$

Correct, 1/1 scored.

(b) Explain the restriction $0 \leqslant t \leqslant 10$. **[3]**

When $t = 10$, $s = 0$ so after 10 s the bird is back at the nest.

Correct; $s = \frac{5t}{2}(10 - t)$ so $s = 0 \rightarrow t = 0$ or 10. The s–t graph shows that s is negative outside the range $0 \leqslant t \leqslant 10$, 2/3 scored.

Correct, 2/2 scored.

(c) Find the velocity of the bird at time t seconds. **[2]**

$$v = \frac{ds}{dt} = 25 - 5t$$

Correct – from the graph above, the maximum point occurs, by symmetry, at $t = 5$, 3/3 scored.

(d) Find the greatest distance of the bird from the nest. **[3]**

Greatest distance occurs when $v = 0$
i.e. $25 - 5t = 0 \rightarrow t = 5$
$$s = 25 \times 5 - \frac{125}{2} = 62.5 \text{ m}$$

Correct – the bird is back at the nest after 10 s.

A refined model is then proposed where s is given by

$$s = 10t^2 - 2t^3 + \frac{1}{10}t^4.$$

(e) Show that the two models agree about the time of the journey and the greatest distance from the nest. **[6]**

When $t = 10$, $s = 1000 - 2000 + 1000 = 0$
$v = \frac{ds}{dt} = 20t - 6t^2 + \frac{2}{5}t^3$; *when $t = 5$, $v = 100 - 150 + 50 = 0$*
When $t = 5$, $s = 250 - 250 + 62.5 = 62.5$ m

Partly correct – although the candidate has assumed that the greatest distance will occur at the same time, $t = 5$. Strictly speaking we should solve $v = 0$ i.e. $20t - 6t^2 + \frac{2}{5}t^3 = 0$; multiplying through by 5, dividing by 2, and factorising gives $t(t - 5)(t - 10) = 0$; thus $v = 0$ when $t = 0$, 5 or 10, 5/6 scored.

(f) Compare the predictions of the two models about velocity and suggest why the refined model is better. **[3]**

$$v = \frac{ds}{dt} = 20t - 6t^2 + \frac{2}{5}t^3$$

The candidate is unable to proceed beyond this. From the working above we see that, under the refined model the bird starts off at rest, gradually increasing its speed and then gradually decelerates and lands back at the nest at rest – under the initial model, the bird 'suddenly' starts with a speed of 25 m s⁻¹ and 'crashes' into the nest with a speed of 25 m s⁻¹, 1/3 scored.

Exam practice questions

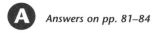 *Answers on pp. 81–84*

7.1 Kinematics (vectors)

(1) A particle P of mass 1.5 kg moves under the action of a single force **F** Newtons. The position vector **r** metres of P at time t seconds relative to a fixed origin O is given by

$$\mathbf{r} = (3t^2 + 4)\mathbf{i} + (4t - t^2)\mathbf{j}$$

where **i** and **j** are perpendicular unit vectors. Find:

(a) the speed of P when $t = 4$ **[5]**

(b) the angle between the direction of motion of P and the vector **i** when $t = 4$ **[2]**

(c) the magnitude of **F**. **[4]**

(2) The velocity **v** m s^{-1} of a particle P at time t seconds is given by

$$\mathbf{v} = (3t^2 - 12)\mathbf{i} + 5t\mathbf{j}$$

(a) Find the acceleration of P at time t seconds. **[2]**

(b) Find the value of t when P is moving parallel to the vector **j**. **[3]**

With respect to a fixed origin O, the position vector of P when $t = 2$ is 4**j**.

(c) Find the distance OP when $t = 1$. **[4]**

7.2 Centres of mass

(1)

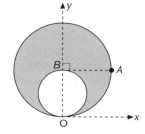

A uniform circular card has radius 6 cm. A circular hole of radius 3 cm is punched in the card in such a way that O lies on the edge of the hole as shown in the diagram. Using the axes shown,

(a) write down the x-coordinate of the centre of mass, G, of the remaining card, giving a reason for your answer, **[2]**

(b) find the y-coordinate of G. **[4]**

The remaining card is now suspended from the point A and hangs in equilibrium.

(c) Find, to the nearest degree, the angle between BA and the horizontal. **[4]**

(2)

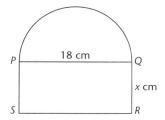

The diagram shows a uniform lamina. $PQRS$ is a rectangle which measures 18 cm by x cm. A semicircle of radius 9 cm is attached to the rectangle with the diameter of the semicircle coinciding with PQ as shown. The centre of mass of the composite lamina lies on PQ. Find the value of x. **[10]**

7.3 Work, energy and power

(1) A small stone of mass 0.5 kg is thrown vertically upwards with an initial speed of 25 m s^{-1}. The maximum height reached by the stone above the level of projection is 20 m.

    (a) Find the energy lost due to air resistance. **[3]**

    (b) The air resistance is modelled as being constant, find its value. **[3]**

(2) A boy throws a stone from the edge of a vertical cliff at a height of 25 m above sea level. The stone is thrown out from the cliff with a speed of 20 m s^{-1} at an angle of 35° above the horizontal. By modelling the stone as a particle and ignoring air resistance, find, correct to one decimal place, the speed of the stone at the instant when it is 15 m above sea level. **[5]**

(3) The non-gravitational resistances opposing the motion of a lorry of mass 2200 kg are constant and total 3100 N.

    (a) The lorry is moving along a straight horizontal road at a constant speed of 54 km h^{-1}. Find the power being developed by the engine of the lorry. **[6]**

The power is suddenly decreased by 16.5 kW.

    (b) Find, in m s^{-2}, the immediate retardation of the lorry. **[3]**

The lorry moves up a hill inclined at an angle θ to the horizontal, where $\sin\theta = \frac{10}{49}$, with the engine working at 90 kW.

    (c) Find, in km h^{-1}, the greatest steady speed which the lorry can maintain up this hill. **[5]**

Exam practice questions

7.4 Collisions

(1) A particle P of mass $2m$, moving with speed $2u$ in a straight line on a smooth horizontal table, collides with another particle Q of mass $3m$, moving with speed u in the same direction as P. The coefficient of restitution between P and Q is e.

  (a) Show that the speed of Q after the collision is $\frac{1}{5}u(7 + 2e)$. **[6]**

  (b) Find, in terms of u and e, the speed of P after the collision. **[2]**

  Given that the speed of P after the collision is $\frac{11}{10}u$,

  (c) show that $e = \frac{1}{2}$. **[2]**

  At the instant of collision, P and Q are a distance d from a fixed vertical wall which is perpendicular to their direction of motion. Given that Q hits the wall, and that the coefficient of restitution between Q and the wall is $\frac{11}{16}$,

  (d) find the distance of P from the wall at the instant that Q hits the wall, **[4]**

  (e) show that, after Q rebounds from the wall, it collides with P again at a distance $\frac{5}{32}d$ from the wall. **[4]**

(2) A small ball A, of mass 120 g, is moving with speed 14 m s^{-1}. It collides directly with another small ball B, of mass 100 g, moving with speed 16 m s^{-1} in the opposite direction. The coefficient of restitution between A and B is e. Immediately after the collision the speed of B is v m s^{-1}.

  (a) Show that $v = \frac{1}{11}(180e + 4)$. **[6]**

  (b) Show that $\frac{4}{11} \leqslant v \leqslant 16\frac{8}{11}$. **[3]**

  (c) Find, in terms of e, the velocity of A. **[2]**

  Given that A is brought to rest by the collision,

  (d) find the value of e. **[6]**

7.5 Statics of rigid bodies

(1) A uniform rod PQ, of mass m, rests in equilibrium with the end P on rough horizontal ground and the end Q against a smooth vertical wall. The vertical plane containing the rod is perpendicular to the wall. The rod is inclined at an angle α to the vertical and the coefficient of friction between the rod and the ground is 0.2.

Show that $\tan\alpha \leqslant 0.4$.

Answers

7.1 Kinematics (vectors)

(1) (a) $\mathbf{v} = 6t\mathbf{i} + (4 - 2t)\mathbf{j}$

When $t = 4$, $\mathbf{v} = 24\mathbf{i} - 4\mathbf{j}$

Speed, $|\mathbf{v}| = \sqrt{(24^2 + 4^2)} = \sqrt{590} = 24.3 \text{ m s}^{-1}$.

Differentiate $\mathbf{r}$ to get the velocity vector.
Speed is the magnitude of the velocity.

(b)

$\tan\alpha = \frac{4}{24}$

$\alpha = 9.46°$.

The direction of motion is given by the direction of $\mathbf{v}$.
It is always worth drawing a simple diagram.

(c) $\mathbf{a} = 6\mathbf{i} - 2\mathbf{j}$

$\mathbf{F} = 1.5\mathbf{a} = 9\mathbf{i} - 3\mathbf{j}$

$|\mathbf{F}| = \sqrt{(9^2 + 3^2)} = \sqrt{90} = 9.49 \text{ N (3 s.f.)}$.

Differentiate $\mathbf{v}$ to get the acceleration vector.
Using $\mathbf{F} = m\mathbf{a}$.

(2) (a) $\mathbf{a} = 6t\mathbf{i} + 5\mathbf{j}$.

(b) $(3t^2 - 12) = 0 \rightarrow t^2 - 4 = 0 \rightarrow t = 2$.

$\mathbf{v}$ parallel to $\mathbf{j} \rightarrow \mathbf{i}$ component = 0.

(c) $\mathbf{r} = (t^3 - 12t + c)\mathbf{i} + \left(\frac{5}{2}t^2 + d\right)\mathbf{j}$

When $t = 2$, $\mathbf{r} = 4\mathbf{j}$

i.e. $4\mathbf{j} = (8 - 24 + c)\mathbf{i} + (10 + d)\mathbf{j}$

$(8 - 24 + c) = 0$ and $(10 + d) = 4$

$c = 16$, $d = -6$, so $\mathbf{r} = (t^3 - 12t + 16)\mathbf{i} + \left(\frac{5}{2}t^2 - 6\right)\mathbf{j}$

When $t = 1$, $\mathbf{r} = 5\mathbf{i} - 3.5\mathbf{j}$, so $OP = \sqrt{(25 + 12.25)} = 6.10 \text{ m}$.

Integrating $\mathbf{v}$ to get $\mathbf{r}$; note the two arbitrary constants using the given conditions.
Equating coefficients of $\mathbf{i}$ and $\mathbf{j}$.

7.2 Centres of mass

(1) (a)   By symmetry, $\bar{x} = 0$.

(b)

	large circle	−	small circle	=	remaining card
Relative mass	36π		9π		27π
y-coordinate	6		3		$\bar{y}$
	$36\pi \times 6$	−	$9\pi \times 3$	=	$27\pi y$

So $\bar{y} = 7$.

Setting out the work in tabular form as shown will help to clarify your thoughts. Since the card is uniform the mass will be proportional to the area. Note that only the relative mass is required so a ratio of $4 : 1 : 3$ would have done.

(c)   Required angle is AGB.

$\tan AGB = \frac{6}{1} \rightarrow AGB = 81°$, to the nearest degree.

It is not necessary to draw the card hanging. Just draw a line on your diagram from the point of suspension, A, to the centre of mass, G. This will then be the vertical and allow you to identify which angle is required.

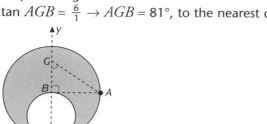

Answers

(2)

	rectangle	+	semicircle	= lamina
Relative mass	$18x$		$\frac{1}{2}\pi.9^2$	$18x + \frac{1}{2}\pi.9^2$.

Distance
from PQ $-\frac{x}{2}$ $\frac{4}{3\pi}.9$ 0

$$18x.\left(-\frac{x}{2}\right) + \frac{1}{2}\pi.9^2\left(\frac{4}{3\pi}\times9\right) = 0$$

$9x^2 = 6\ 9^2$
$x^2 = 54$
$x = 3\sqrt{6}. = 7.35$ cm.

Note that since the centres of mass of the rectangle and semicircle are on opposite sides of PQ the distances must be given opposite signs. To avoid this you could measure all your distances from SR, for example. The formula for the position of the centre of mass of a semicircle of radius r is $\frac{4r}{3\pi}$ from the centre and can be taken out of the formula booklet.

7.3 Work, energy and power

(1) (a)
Energy loss = Losses − Gains
= KE Loss − PE Gain
= $\frac{1}{2}0.5\ 25^2 - 0.5g.20$
= 58.25 J.

It is worth writing out the first line in full and then decide which type of energy is being lost and which type of energy is being gained.

(b)
Work done against a resistance
= Overall loss of energy
$20 \times R = 58.25$
$R = 2.9125$ N.

Note that we never include work done against gravity since this is allowed for in the PE term.

(2)
Since there is no air resistance, energy is conserved.
KE Gain = PE Loss
$\frac{1}{2}mv^2 - \frac{1}{2}m.20^2 = mg.10$
$v^2 = 20^2 + 20g = 596$
$v = 24.4$ m s^{-1}.

At first sight this appears to be a projectiles question, and of course, it could be solved by finding the horizontal and vertical components of velocity, and using these to find the speed, but the Conservation of Energy Principle provides a much quicker and neater solution. Note also that it is not necessary to find how far up the stone goes – this is the beauty of using energy – since the total energy of the stone is the same at all positions, there is no need, or indeed point, in considering intermediate positions.

(3) (a)
54 km h^{-1} = $\frac{54\,000\,m}{3600\,s}$ = 15 m s^{-1}
Constant speed → zero acceleration →
$D = 3100$, where D is the driving force.
$P = Fv = 3100 \times 15 = 46\ 500$ W = 46.5 kW.

(b)
New power is 30 000 W
Driving force = $\frac{30\,000}{15}$ = 2000 N
Resolving gives: 2000 − 3100 = 2200a
$a = -\frac{1}{2}$ m s^{-2}
So retardation is $\frac{1}{2}$ m s^{-2}.

Speeds must be measured in m s^{-1}, for the equation $P = Fv$ to be valid. If there is zero acceleration, the driving force must equal the resistance.

(c)
Resolving up the hill,
$D - 2200g\sin\theta - 3100 = 0 \rightarrow D = 7500$ N
Using $P = Fv$,
$90\ 000 = 7500v \rightarrow v = 12$ m s^{-1}
= $12 \times \frac{3600}{1000}$ km h^{-1} = 43.2 km h^{-1}.

In any dynamics problem, always resolve in the direction of motion.

It may be worth drawing a diagram to try to ensure that you don't omit any of the forces, particularly the weight of the lorry.

7.4 Collisions

(1) (a) Always draw a simple diagram as follows:

$\longrightarrow 2u \quad \longrightarrow u$

$\left(2m\right) \quad e \quad \left(3m\right)$

$\longrightarrow v_1 \quad \longrightarrow v_2$

$$2m2u + 3mu = 2mv_1 + 3mv_2$$
$$7u = 2v_1 + 3v_2 \qquad (1)$$
$$e(2u - u) = -v_1 + v_2$$
$$eu = -v_1 + v_2 \qquad (2)$$
$$2eu = -2v_1 + 2v_2 \qquad 2 \times (2)$$
$$7u = 2v_1 + 3v_2 \qquad (1)$$

Adding gives
$$u(7 + 2e) = 5v_2 \rightarrow v_2 = \tfrac{u}{5}(7 + 2e)$$

(b)
$$eu = -v_1 + v_2 \qquad (2)$$
$$3eu = -3v_1 + 3v_2 \qquad 3 \times (2)$$
$$7u = 2v_1 + 3v_2 \qquad (1)$$

Subtracting top from bottom gives,
$$7u - 3ue = 5v_1 \rightarrow v_1 = \tfrac{u}{5}(7 - 3e) > 0 \text{ as } 3e < 7,$$
so this is a speed.

(c) $\quad \frac{11u}{10} = \frac{u}{5}(7 - 3e) \rightarrow 11 = 14 - 6e \rightarrow e = \frac{1}{2}$

(d) When $e = \frac{1}{2}$, $v_2 = \frac{u}{5}(7 + 1) = \frac{8u}{5}$

Time for Q to hit the wall $= d/\left(\frac{8u}{5}\right) = \frac{5d}{8u}$

In this time P travels $\frac{11u}{10} \times \frac{5d}{8u} = \frac{11d}{16}$

So P is $\frac{5d}{16}$ from the wall.

(e) Speed of Q after rebound $= \frac{11}{16} \times \frac{8u}{5} = \frac{11u}{10}$

So P and Q are now travelling with the same speed $\left(\frac{11u}{10}\right)$ towards each other.

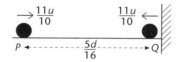

$\longrightarrow \frac{11u}{10} \qquad\qquad \frac{11u}{10} \longleftarrow$

$P \longleftarrow \text{------}\frac{5d}{16}\text{------}\longrightarrow Q$

Hence they will collide at a point which is $\frac{1}{2} \times \frac{5d}{16}$ from the wall, i.e. $\frac{5d}{32}$.

Examiner's tips

We must now solve (1) and (2) to find v_2. Write the two simultaneous equations with 'unknowns' on one side, underneath each other, and 'knowns' (i.e. things given in the question) on the other side, and then solve by elimination.

As required.

You must always check, if you are asked to find a speed, that the answer that you have given is definitely positive – sometimes this may depend on some restriction on e that is given in the question, e.g. you may be told that $e < \frac{1}{2}$, so if you end up with $v = u(2e - 1)$ and the question requires a speed, then you have chosen the wrong direction for v and your answer should be changed to $u(1 - 2e)$, which is positive since $e < \frac{1}{2}$.

It is always worth drawing a simple diagram.

Answers

(2) (a)

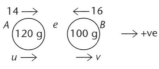

Draw a diagram showing the collision. Mark your positive direction on the diagram.

Conservation of momentum:
$$120 \times 14 - 100 \times 16 = 120u + 100v$$
$$80 = 120u + 100v$$
$$4 = 6u + 5v \qquad (1)$$

Newton's Law:
$$v - u = e(16 + 14) = 30e$$
Rewrite the equations as:
$$5v + 6u = 4 \qquad (1)$$
$$v - u = 30e \qquad (2)$$
$$6v - 6u = 180e \qquad 6 \times (2)$$
$$11v = 180e + 4$$
$$v = \tfrac{1}{11}(180e + 4).$$

Note that when writing down the momentum equation, there is no need to have the masses in kg – provided all the masses are in the same units, the equation will be correct – but you must choose a positive direction. Sign errors are the most common mistakes made in this type of question.

To eliminate u multiply (2) by 6. Now add to (1).

(b) From part **(a)**, $e = \tfrac{1}{180}(11v - 4)$

Now, $0 \leqslant e \leqslant 1$, so $0 \leqslant \tfrac{1}{180}(11v - 4) \leqslant 1$
$$0 \leqslant (11v - 4) \leqslant 180$$
$$4 \leqslant 11v \leqslant 184$$
$$\tfrac{4}{11} \leqslant v \leqslant \tfrac{184}{11} = 16\tfrac{8}{11}.$$

When asked to prove or derive an inequality, you must start with an inequality, not an equation.

(c) From above,
$$5v + 6u = 4 \qquad (1)$$
$$v - u = 30e \qquad (2)$$
$$5v - 5u = 150e \qquad 5 \times (2)$$
$$11u = (4 - 150e)$$
$$u = \tfrac{1}{11}(4 - 150e), \text{ in original direction.}$$

Subtracting from (1), to eliminate v.

Note that this is a velocity, not a speed.

(d)
$$u = 0 \rightarrow \tfrac{1}{11}(4 - 150e) = 0$$
$$\rightarrow \quad (4 - 150e) = 0$$
$$\rightarrow \qquad\qquad e = \tfrac{4}{150} = \tfrac{2}{75}.$$

7.5 Statics of rigid bodies

(1)

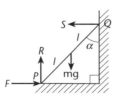

First draw a large clear diagram showing all the forces.

Let the length of the rod be $2l$.
$R(\rightarrow)$, $F = S$
$R(\uparrow)$, $R = mg$
$M(P)$, $mgl\sin\alpha = S.2l\cos\alpha$
$\rightarrow S = 0.5\, mg\tan\alpha$ i.e. $F = 0.5\, mg\tan\alpha$
Now, $F \leqslant \mu R \rightarrow 0.5\, mg\tan\alpha \leqslant 0.2\, mg \rightarrow \tan\alpha \leqslant 0.4$ as required.

If the question states that 'the body is in limiting equilibrium' or 'on the point of slipping' or similar, then you should put μR on your diagram for the friction force, but otherwise just use F on the diagram and then use, in your calculations, the fact that $F \leqslant \mu R$, to derive any required inequality.

Questions with model answers

? *For help see Revise A2 Study Guide pages 100 and 101 and 51–53*

C grade candidate – mark scored 5/9

(1) A balloon manufacturer makes the following claim: '95% of our balloons will not burst when they are inflated'. There are 20 balloons to be blown up for a Christmas party.

(a) What is the probability that

    (i)   none of them burst,   **[2]**

$$Prob = 0.95^{20} = 0.358 \ (3 \ s.f.)$$

> *Examiner's Commentary*
>
> *Correct – this is a Binomial distribution, **2/2 scored**.*

    (ii)   exactly two of them burst?   **[3]**

$$Prob = {}^{20}C_2 0.95^{18}. \ 0.05^2 = 0.189 \ (3 \ s.f.)$$

> *Correct also, **3/3 scored**.*

The balloons are sold in packets of 20 and a packet for which none of its balloons burst when inflated is known as a Star Packet.

A child goes into a shop and buys 10 packets.

(b) Find the probability that half of them are Star Packets.   **[4]**

$$Prob = 0.5 \times 0.358 = 0.179$$

> *Incorrect – although the candidate has realised that he needs to use the answer to part **(a)(i)**, he has not realised that we have another Binomial distribution – if S is the no. of Star Packets, then S ~ B(10, 0.358), so P(S = 5) = ${}^{10}C_5.0.358^5.0.642^5$ = 0.162 (3 s.f.), **0/4 scored**.*

A grade candidate – mark scored 15/17

(2) The lifetime, T years, of a brand of electric light bulb is modelled by the probability density function, $f(t)$, where

$$f(t) = kt \, (5 - t), \quad \text{for } 0 \leqslant t \leqslant 5, \text{ where } k \text{ is a positive constant}$$
$$ = 0 \qquad\qquad \text{otherwise}$$

(a) Show that $k = \frac{6}{125}$.   **[3]**

$$\int_0^5 kt \, (5 - t) \, dt = 1$$

$$k \int_0^5 5t - t^2 \, dt = 1$$

$$k \left[\frac{5}{2} t^2 - \frac{1}{3} t^3 \right]_0^5 = 1 \rightarrow k = \frac{6}{125}$$

> *Correct; the total area under any p.d.f. is 1. Leaving the constant k outside the integral is a good idea. **3/3 scored**.*

Questions with model answers

For help see Revise A2 Study Guide pages 51–53

A grade candidate continued

(b) Sketch $f(t)$ for all values of t. **[3]**

> $f(t)$ is a quadratic in t and so its graph will be a parabola.
> Solve $f(t) = 0$ to find where it cuts the horizontal axis.
> $\frac{6}{125}t(5-t) = 0 \rightarrow t = 0$ or 5

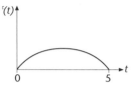

Examiner's Commentary

> The candidate loses a mark as she hasn't made it clear what happens for t outside $0 \leqslant t \leqslant 5$, **2/3 scored.**

(c) Find $E[T]$. **[2]**

$$E[T] = \int_0^5 kt^2(5-t)dt = \frac{6}{125}\int_0^5 5t^2 - t^3\, dt$$

$$= \frac{6}{125}\left[\frac{5}{3}t^3 - \frac{1}{4}t^4\right]_0^5$$

$$= 2.5$$

> All correct but the candidate has wasted valuable time and effort, since from the graph, by symmetry, $E(T) = 2.5$, **2/2 scored.**

(d) A standard lamp has just been fitted with a new bulb. Find the probability that the bulb will fail within 2 years. **[3]**

$$P(t \leqslant 2) = \int_0^2 kt(5-t)dt = \frac{6}{125}\left[\frac{5}{2}t^2 - \frac{1}{3}t^3\right] = \frac{44}{125}$$

> Correct, **3/3 scored.**

(e) Find the standard deviation of T. **[5]**

$$Var\,(T) = \int_0^5 kt^3(5-t)\,dt = \frac{6}{125}\left[\frac{5}{4}t^4 - \frac{1}{5}t^5\right]_0^5 = 7.5$$

$$so,\ S.D. = \sqrt{7.5} = 2.74$$

> Incorrect – the candidate has forgotten to subtract μ^2 – given the graph of f, the candidate should have realised that this value is far too big.

(f) Write down fully the cumulative distribution function of T. **[2]**

$$F(t) = \begin{array}{ll} 0 & \text{for } t < 0 \\ \frac{6}{125}\left(\frac{5}{2}t^2 - \frac{1}{3}t^3\right) & \text{for } 0 \leqslant t \leqslant 5 \\ 1 & \text{for } t > 5 \end{array}$$

> Correct, **2/2 scored.**

Exam practice questions

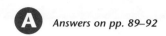

Answers on pp. 89–92

8.1 Binomial and Poisson distribution

(1) The number, B, of breakdowns per week of the lifts in a large block of flats has a Poisson distribution with a mean of 0.25. Find, to three decimal places, the probability that in a particular week:

(a) there will be at least one breakdown **[2]**

(b) there will be at most two breakdowns. **[3]**

(c) Show that, to three decimal places, the probability that, during a 12 week period, there will be no lift breakdowns is 0.050. **[5]**

The residents association of the block of flats has a maintenance contract with *Liftfix*. The contract is for 20 periods of 12 weeks. For every period of 12 weeks without a breakdown the association pays *Liftfix* £500. If there is at least one breakdown during a 12 week period then *Liftfix* will fix the lift free of charge and the association pays nothing for that period.

(d) Find the probability that, during the time of the contract, the association pays no more than £1000. **[3]**

(2) For each call into the telephone exchange of a large company, there is an independent probability of 0.002 that it is connected to the wrong extension.

(a) Find, to 3 significant figures, the probability that, on a given day, exactly one of the first five incoming calls will be wrongly connected. **[4]**

On one particular day the company receives 1000 calls.

(b) Use a Poisson approximation to find, to 3 decimal places, the probability that at least three of them are wrongly connected. **[4]**

8.2 Continuous random variables

(1) The cumulative distribution function of a random variable X is given by

$$F(x) = \begin{cases} 0, & 1 > x \\ \frac{1}{8}(x^2 - 1), & 1 \leqslant x \leqslant 3 \\ 1, & 3 < x \end{cases}$$

(a) Find the probability density function, f, of X. **[3]**

(b) Sketch the graph of f. **[2]**

(c) Find the mode of X. **[3]**

(d) Find the median of X. **[4]**

(e) Find the mean of X. **[4]**

(f) Describe the distribution of X. **[2]**

Exam practice questions

8.3 Continuous distributions

(1) In a multiple-choice examination, there is only one correct answer to each question, and for each question there are four possible answers given, from which the candidates select one. A candidate answering 20 or more questions correctly, passes the examination.

A particular candidate, I.M. Clueless, decides to select his answers to each question at random.

(a) Using a Normal approximation, find, to 2 significant figures, the probability that Clueless passes the examination if it contains 50 questions. **[6]**

(b) If it is required that the probability that Clueless passes is less than 0.005, find the largest number of questions that there should be on the paper. **[3]**

(2) A company, 4U2Hire, hires out vans on a daily basis. The mean number of vans hired per day is 15. Using the normal approximation for a Poisson distribution, find, for a period of 100 days:

(a) how often five or less vans are hired out in a day, **[6]**

(b) how often exactly 10 vans are hired out in a day, **[3]**

(c) on how many days they will have to turn customers away if the company owns 20 vans. **[3]**

8.4 Hypothesis tests

(1) The probability of Skinclear lotion curing a particular skin complaint is 0.35. It is thought that a new product, Cureall lotion, will be more effective and improve the cure rate. The new product was tested on 20 people; 10 people found it worked and cured their problem.

(a) Stating your null and alternative hypotheses carefully, show that the cure rate of the Cureall lotion is not significant at the 5% level. **[10]**

(b) How many of the 20 people tested would need to be cured for the result to be significant at the 5% level? **[3]**

(2) In a certain restaurant it is believed that the probability that a customer will order a vegetarian meal is 0.1. In order to try to ensure that their provision is appropriate, the manager decides to take a random sample of 100 customers to test to see if the proportion requiring vegetarian meals is different to 0.1. By using a suitable approximation, find the critical region for this test at the 5% level of significance. **[4]**

The manager finds that 19 out of the sample of 100 customers ask for a vegetarian meal; what conclusions could he draw from this? **[2]**

Answers

8.1 Binomial and Poisson distribution

(1) (a) $B \sim Po(0.25)$

$P(B \geqslant 1) = 1 - P(B = 0) = 1 - e^{-0.25} = 0.221$ (3 d.p.).

(b) $P(B \leqslant 2) = P(B = 0) + P(B = 1) + P(B = 2)$

$= e^{-0.25} \left(1 + 0.25 + \frac{(0.25)^2}{2!} \right)$

$= 0.998$ (3 d.p.)

Let M be the number of breakdowns in a 12 week period.

Then $M \sim Po(12 \times 0.25) = Po(3)$.

(c) $P(M = 0) = e^{-3} = 0.04978... = 0.050$ (3 d.p.).

(d) Let X be the number of periods (out of the 20 under the contract) in which there are no breakdowns.

Then $X \sim B(20, 0.05)$. For the association to pay out $\leqslant £1000$, we need $X \leqslant 2$.

$P(X \leqslant 2) = P(X = 0) + P(X = 1) + P(X = 2) = 0.95^{20}$
$+ {}^{20}C_1 0.05 \times 0.95^{19} + {}^{20}C_2 0.05^2 \times 0.95^{18} = 0.9245$.

(2) (a) Let X be the number of calls (of the first 5) that are wrongly connected.

Then $X \sim B(5, 0.002)$

$P(X = 1) = {}^5C_1 0.002 \times 0.998^4 = 0.00992$ (3 s.f.).

(b) Let Y be the number of calls (of the 1000) that are wrongly connected.

Then $Y \sim Bin(1000, 0.002)$.

Since n is large and p is small then an approximation for Y is given by $Y \sim Po(1000 \times 0.002) = Po(2)$

We require $P(Y \geqslant 3)$

$= 1 - \{P(Y = 0) + P(Y = 1) + P(Y = 2)\}$

$= 1 - e^{-2} \left(1 + 2 + \frac{2^2}{2!} \right)$

$= 1 - 5e^{-2}$

$= 0.32332358...$

$= 0.323$ (3 d.p.).

Note that we could have found the exact value of $P(Y \geqslant 3)$ using $B(1000, 0.002)$, as follows:
$P(Y \geqslant 3) = 1 - \{P(Y = 0) +$
$P(Y = 1) + P(Y = 2)\}$
$= 0.998^{1000} + {}^{1000}C_1 0.002 \times$
$0.998^{999} + {}^{1000}C_2 0.002^2 \times 0.998^{998}$
$= 0.32332349...$
In this case the two figures agree to 6 d.p. since n is very large and p is very small.

8.2 Continuous random variables

(1) (a) $f(x) = F'(x) = \frac{x}{4}$

$$f(x) = \begin{cases} 0, & 1 > x \\ \frac{x}{4}, & 1 \leqslant x \leqslant 3 \\ 0, & 3 < x. \end{cases}$$

Answers

(b)

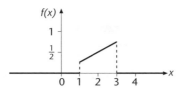

The mode is the value of X for which $f(x)$ is the greatest. Sometimes it may be necessary to use $f'(x)$ to find the mode, by solving $f'(x) = 0$ to find a maximum point on the graph of $f(x)$.

(c) From the graph, the mode of X is 3.

(d) $F(m) = 0.5 \rightarrow \frac{1}{8}(m^2 - 1) = 0.5$

$ \rightarrow \qquad m^2 = 5$

$ \rightarrow \qquad m = \sqrt{5}$

There is no need to do any integration here since we already have the c.d.f. Note that the lower quartile, q_1, is given by $F(q_1) = 0.25$ and the upper quartile, q_2, is given by $F(q_2) = 0.75$.

(e) $\mu = \int_1^3 x\,f(x)\,\mathrm{d}x = \frac{1}{4}\int_1^3 x^2\mathrm{d}x = \frac{1}{12}\left[x^3\right]_1^3 = \frac{1}{12}(27 - 1) = \frac{13}{6}$

(f) Here, mean < median < mode.

X is negatively skewed as can be seen from the graph of its density function.

8.3 Continuous distributions

(1) (a) Let X be the number (out of 50) of correct answers obtained.

Then $X \sim B(50, 0.25)$.

We require $P(X \geqslant 20) = 1 - P(X \leqslant 19)$.

We can approximate X by $N(50 \times 0.25,\ 50 \times 0.25 \times 0.75) = N(12.5, 9.375)$

Since we are approximating a discrete distribution

by a continuous one, we require a continuity correction,

i.e. $1 - P(X \leqslant 19.5) = 1 - P(Z \leqslant (19.5 - 12.5)/\sqrt{9.375})$

$ = 1 - P(Z \leqslant 2.286)$

$ = 1 - \Phi(2.286)$

$ = 1 - 0.9889$

$ = 0.011\ (2\ \text{s.f.})$

When deciding on the appropriate continuity correction, it is a good idea to draw a diagram.

(b) Suppose there are n questions on the paper.
Let Y be the number (out of n) of correct answers obtained.
Then $Y \sim B(n, 0.25)$.
We require $P(Y \geq 20) < 0.005$
i.e. $\quad 1 - P(Y \leq 19) < 0.005$
i.e. $\quad\quad\quad 0.995 < P(Y \leq 19)$
Now we can approximate Y by $N\left(\frac{n}{4}, \frac{3n}{16}\right)$, and using a continuity correction,

$$0.995 < P(Y \leq 19.5) = P\left(Z \leq \left(19.5 - \frac{n}{4}\right)/\sqrt{\left(\frac{3n}{16}\right)}\right)$$

$$\rightarrow \Phi\left(19.5 - \frac{n}{4}\right)/\sqrt{\left(\frac{3n}{16}\right)} > 0.995$$

$$\rightarrow \left(19.5 - \frac{n}{4}\right)/\sqrt{\left(\frac{3n}{16}\right)} > 2.58 \text{ (from tables using interpolation)}$$

$$\rightarrow \quad\quad (78 - n) > 2.58 \times \sqrt{(3n)}$$

$$\rightarrow \quad\quad\quad 0 > n + 4.4687\sqrt{n} - 78$$

Let $x = \sqrt{n}$, i.e. $x > 0$
$$0 > x^2 + 4.4687x - 78$$

Using the quadratic formula, $-11.344 < x < 6.8756$ but $x > 0$
So, $0 < x < 6.8756$ i.e. $0 < \sqrt{n} < 6.8756 \rightarrow n > 47.27...$
So the maximum number of questions is 47.

(2) (a) Let V be the number of vans hired per day.
Then $V \sim Po(15)$.
We can approximate V using $N(15, 15)$.
We require $P(V \leq 5)$, but since we are approximating a discrete distribution by a continuous one we need a continuity correction.
$P(V \leq 5)$ becomes $P(V \leq 5.5)$
$= P(Z \leq (5.5 - 15)/\sqrt{15})$
$= P(Z \leq -2.4529)$
$= 1 - P(Z \leq 2.4529)$
$= 1 - \Phi(2.4529)$
$= 0.0070$
Hence in a 100 day period we would expect five or less vans to be hired on 0.70 days, i.e. on 1 day.

(b) $P(V = 10) = P(9.5 \leq V \leq 10.5)$
$= P((9.5 - 15)/\sqrt{15} \leq Z \leq (10.5 - 15)/\sqrt{15})$
$= P(-1.4201 \leq Z \leq -1.1619)$
$= P(1.16 \leq Z \leq 1.42)$
$= \Phi(1.42) - \Phi(1.16)$
$= 0.0448$
In a 100 day period this becomes 4.48
i.e. on 4 days.

(c) A customer will have to be turned away if $V > 20$

so we require $P(V > 20) = 1 - P(V \leqslant 20)$

$= 1 - P(V \leqslant 20.5)$, with the continuity correction

$= 1 - P(Z \leqslant (20.5 - 15)/\sqrt{15})$

$= 1 - P(Z \leqslant 1.4201)$

$= 1 - \Phi(1.42)$

$= 0.0778$

In a 100 day period this becomes 7.78 i.e. on 8 days.

8.4 Hypothesis tests

(1) (a) Let X be the number of patients out of 20 who are cured by the new product.

$H_0: p = 0.35$ $H_1: p > 0.35$

Assuming H_0 is true then $X \sim B(20, 0.35)$

$P(X \geqslant 10) = 1 - P(X \leqslant 9)$

$= 1 - 0.8782$, using the cumulative Binomial probability tables in the formula booklet

$= 0.1218 > 0.05$

Hence the test result is not significant at the 5% level and there is insufficient evidence to reject the null hypothesis.

(b) With the same notation and hypotheses as above, we need to find c such that

$P(X \geqslant c) \leqslant 0.05$

$1 - P(X \leqslant c - 1) \leqslant 0.05$

$P(X \leqslant c - 1) \geqslant 0.95$

$c - 1 = 11$, using the cumulative Binomial probability tables in the formula booklet

$c = 12$.

(2) Let X be the number of customers, out of 100, who choose a vegetarian meal.

$H_0: p = 0.1$ $H_1: p \neq 0.1$

Under H_0, $X \sim B(100, 0.1)$. Since n is large and p is small, $X \sim Po(100 \times 0.1)$ is a suitable approximation, i.e. $X \sim Po(10)$.

For a two-tailed test we require c_1 and c_2 such that

$P(X \leqslant c_1) \leqslant 0.025$ and $P(X \geqslant c_2) \leqslant 0.025$

$P(X \leqslant c_1) \leqslant 0.025 \rightarrow c_1 = 3$, using the cumulative Poisson probability tables

and $P(X \geqslant c_2) \leqslant 0.025 \rightarrow 1 - P(X \leqslant c_2 - 1) \leqslant 0.025$

$\rightarrow P(X \leqslant c_2 - 1) \geqslant 0.975$

$\rightarrow c_2 - 1 = 17$

$\rightarrow c_2 = 18$

So critical region is $X \leqslant 3$ and $X \geqslant 18$.

Since 19 is in the critical region, there is sufficient evidence at the 5% level to reject H_0.

Questions with model answers

For help see Revise AS Study Guide pages 114 and 115

C grade candidate – mark scored 9/14

Examiner's Commentary

(1) A cyclist wishes to use some specially prepared cycle tracks to visit five towns. The table gives the distances between the towns in kilometres.

	A	B	C	D	E
A	–	4	5	3	6
B	4	–	8	2	1
C	5	8	–	7	4
D	3	2	7	–	9
E	6	1	4	9	–

(a) Use Prim's algorithm to find a minimum connector for the network. Give the connectors and the total length. **[3]**

Starting at E, connect E–B   1
Connect B–D   2
Connect D–A   3
Connect E–C   4

The total length is omitted, 2/3 marks.

(b) A cyclist wishes to visit each town in the network at least once, and then return to the start. By using each edge of your minimum connector twice, produce a route for the cyclist, starting and finishing at A. Give the length of this route. **[3]**

From A visit each town and return to A: giving 20 km

This is an ambiguous answer, though the 20 km is correct. The route has to be specified, 2/3 marks.

(c) Using the reduced network technique, give a lower bound for a tour of all the towns. **[3]**

e.g. remove A, then the minimum connection gives 9 with
A connected via C and E with 6, giving 15

Correct 3/3 marks.

(d) The cyclist wishes to visit each town exactly once, and then return to the start. Use a greedy algorithm to find such a tour. Start from A. Give the tour and its length. Explain your working. **[5]**

Start at A;
Travel to D   3 km
Travel to B   2 km
Travel to E   1 km
Travel to C   4 km
Return to A   5 km
Total distance of tour   15 km

We would expect the candidate to explain why the cyclist chooses D first. This is the 'greedy algorithm bit' – otherwise the choice may be guesswork, 3/5 marks.

Questions with model answers

For help see Revise AS Study Guide pages 122 and 124

A grade candidate – mark scored 14/16

(2) A factory is set up to produce three types of units, X, Y, Z, and can sell them for £8, £12 and £10 respectively. Restrictions on machinery mean that the total number of Y and Z produced per day cannot exceed 30. X, Y and Z take 4 man hours, 3 man hours and 5 man hours respectively, and the total number of man hours per day available is 150.

Examiner's Commentary

(a) Set up the objective function, if x, y and z items of each unit are to be made per day. **[1]**

If P is the income from sales, then we wish to maximise the income. Given that the income is £8 for each X, £12 for each Y and £10 for each Z, then let there be x, y, z items of each produced per day. Then the income is given by $P = 8x + 12y + 10z$.

Well explained, 1/1 mark.

(b) Explain how the constraint $y + z \leqslant 30$ is produced and find the other constraint. **[3]**

Total number of units of Y and Z cannot exceed 30, so $y + z \leqslant 30$. The other constraint comes from the number of man-hours available which is 150.
So the other constraint is $4x + 3y + 5z \leqslant 150$.

Again, all correct. You should aim to explain each step carefully, 3/3 marks.

(c) Set up a simplex tableau to solve the problem. **[3]**

P	x	y	z	r	s	rhs
1	–8	–12	–10	0	0	0
0	0	1	1	1	0	30
0	4	3	5	0	1	150

All correct, 3/3 marks.

(d) Perform three iterations of the simplex algorithm. **[6]**

P	x	y	z	r	s	rhs	
1	0	–6	0	0	2	300	$R_1 + 2R_3$
0	0	1	1	1	0	30	
0	4	3	5	0	1	150	

The usual practice is to deal first with the coefficient which is largest (the y coefficient), but this is correct and yields the right answer, 6/6 marks.

P	x	y	z	r	s	rhs	
1	0	0	6	6	2	480	$R_1 + 6R_2$
0	0	1	1	1	0	30	
0	4	0	2	–3	1	60	$R_3 - 3R_2$

(e) Suggest a solution and state whether it is optimal. **[3]**

Because the top line is all positive, this suggests that if we make $z = 0$, then the second line becomes $y = 30$ and the third line becomes $4x = 60$, giving $x = 15$.
So 15 of X should be made and 30 of Y.

The statement whether the solution is optimal is not given – while it may be assumed, the question asks for a specific answer. In addition, the income is not specifically given, 1/3 marks.

Exam practice questions

A *Answers on pp. 97–98*

(1) A two-person zero-sum game is represented by the following pay-off matrix for player A.

		B	
		X	Y
A	X	5	2
	Y	3	7

Find

(a)   the best strategy for each player,   **[7]**

(b)   the value of the game.   **[3]**

(2) A large room is to be prepared for a wedding reception. The tasks that need to be carried out are:

I   clean the room

II   arrange the tables and chairs

III   set the places

IV   arrange the decorations

The tasks need to be completed consecutively and the room must be prepared in the least possible time. The tasks are to be assigned to four teams of workers A, B, C and D. Each team must carry out only one task. The table below shows the times, in minutes, that each team takes to carry out each task.

	A	B	C	D
I	17	24	19	18
II	12	23	16	15
III	16	24	21	18
IV	12	24	18	14

(a)   Use the Hungarian algorithm to determine which team should be assigned to each task. You must make your method clear and show:

   (i)   the state of the table after each stage in the algorithm,

   (ii) the final allocation.   **[11]**

(b)   Obtain the minimum total time taken for the room to be prepared.   **[2]**

Exam practice questions

(3) An area representative for a company needs to visit five factories in a particular area in one day. The table below shows the distances, in km, between the five factories A, B, C, D and E. The route is to be planned, starting and finishing at A, visiting each shop once and covering a minimum distance.

	A	B	C	D	E
A	–	16	27	28	14
B	16	–	15	38	25
C	27	15	–	39	26
D	28	38	39	–	12
E	14	25	26	12	–

(a) Obtain a minimum spanning tree for the network, and draw this tree. Start with A and state the order in which the shops are added. **[5]**

Given that the problem is complete and satisfies the triangle inequality,

(b) determine an initial upper bound for the length of the route travelled by the representative, **[2]**

(c) determine a lower bound for the length of the route. **[4]**

Answers

(1) (a) Suppose A chooses X with probability p (and Y with probability $1 - p$).
Then if B chooses X, the gain for A will be $5p + 3(1 - p)$
And if B chooses Y then the gain for A will be $2p + 7(1 - p)$
Optimal value is then given by $5p + 3(1 - p) = 2p + 7(1 - p) \rightarrow 7p = 4 \rightarrow p = \frac{4}{7}$
So A should choose X with probability $\frac{4}{7}$.
Likewise, if B chooses X with probability q,
then if A chooses X the gain for B will be
$5q + 2(1 - q)$ and if A chooses Y the
gain for B will be $3q + 7(1 - q)$.
Optimal value is given by
$5q + 2(1 - q) = 3q + 7(1 - q) \rightarrow 7q = 5 \rightarrow q = \frac{5}{7}$.
So B should choose X with probability $\frac{5}{7}$.

Examiner's tips

Although you might be able to write down the optimal value straightaway, you are advised to write it all out in full.

(b) Value $= 5p + 3(1 - p) = \frac{20}{7} + \frac{9}{7} = \frac{29}{7}$
or $= 3q + 7(1 - q) = \frac{15}{7} + \frac{14}{7} = \frac{29}{7}$.

Doing both serves as a check.

(2) (a) (i) Reducing the columns gives:

5	1	3	4
0	0	0	1
4	1	5	4
0	1	2	0

Reducing the rows gives:

4	0	2	3
0	0	0	1
3	0	4	3
0	1	2	0

The zeros can be covered by three lines.
Minimum uncovered element is 2.
New matrix becomes:

2	0	0	1
0	2	0	1
1	0	2	1
0	3	2	0

This question asks for the tables to be shown. There is no reason why columns should be reduced before rows and credit will be given either way. It would be helpful to you if you were to write out the matrices carefully even if not asked. When the new matrix is reduced by the minimum uncovered element, remember that those elements covered by two lines need to be increased by the minimum element.

(ii) Requires four lines to cover all zeros. Therefore assignment is possible.
I   C
II A
III B
IV D

(b) Times for this assignment are $19 + 12 + 24 + 14 = 69$ minutes.

Answers

(3) **(a)** Start with A. Go the minimum distance,
i.e. 14 to E.
Proceed similarly to D(12), then from
A to B(16), then from B to C(15).

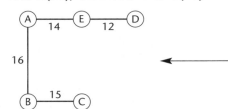

(b) Upper bound = 2 × minimum connector = 257 = 114. ◄———

(c) Remove C. The new minimum connection is 42.
Add to this the least connection of C ◄———
which is CB(15) and CA(27).
This gives 42 + 15 + 27 = 84.

Mock Exam

Time: 1 hour 30 minutes

N.B. Only a basic scientific calculator may be used for this paper

(1) Find the set of values of x for which
$$x^2 > 18 + 7x$$
[4]

(2) The depth of water in a harbour is d metres at time t hours after low tide, where d is modelled by the equation
$$d = 8 - 2\cos ct$$
and c is a positive constant.

(a) Write down the maximum and minimum depth of water in the harbour. **[2]**

Given that, on a particular day, low tide occurs at 10 am and the next high tide occurs at 6 pm,

(b) find, in terms of π, the value of c. **[3]**

(3) $$f(x) = 2x^3 + 6x^2 + 3x - 2.$$

(a) Show that $(x + 2)$ is a factor of $f(x)$. **[2]**

(b) Hence solve $f(x) = 0$, leaving any non-integer answers in the form $a + b\sqrt{c}$, where a, b and c are rational numbers. **[6]**

(4) Solve, for $0 \leqslant x \leqslant 2\pi$,
$$2\sin^2 x = 3 - 3\cos x,$$
giving your answers in terms of π. **[8]**

(5) A manufacturer has a contract to make bricks. Each brick is to be in the shape of a cuboid with base $2x$ cm by x cm and height h cm; the total surface area of each brick must be 300 cm^2.

(a) Show that $h = 50x^{-1} - \frac{2}{3}x$. **[3]**

The volume of the brick is V cm^3.

(b) Find V in terms of x only. **[3]**

Given that x varies,

(c) find the maximum value of V. **[5]**

(6) (a) Prove that, for all real values of x and y,
$$2xy \leqslant x^2 + y^2.$$
[3]

(b) The 1st, 2nd and 3rd terms of an arithmetic series are x^2, A and y^2 respectively.

Find A in terms of x and y. [3]

(c) The 1st, 2nd and 3rd terms of a geometric series are x^2, G and y^2 respectively, where x, G and y are all positive real numbers.

Find G in terms of x and y. [3]

(d) Deduce that $G \leqslant A$. [2]

(e) State the relationship between x and y for which $G = A$. [1]

(7) (a) Show that the line L with equation $y - 2x = 1$ intersects the curve with equation $y^2 = 2x^2 + x$ at the point with coordinates $(-1, -1)$ and find the coordinates of the other point of intersection. [6]

(b) Find the equation of the straight line which is perpendicular to the line L and which passes through the point $(-1, -1)$. [4]

(c) Deduce the distance, to 3 significant figures, of the point with coordinates $(1, -2)$ from the line L. [3]

(8)

The diagram shows part of the curve with equation $y = 3x^2 - 18x + 27$.
The points P and Q lie on the curve and on the line with equation $y = 12$ and R is the point where the curve crosses the x-axis.

(a) Find the coordinates of P and Q. [5]

(b) Find the area of the shaded region bounded by the curve and the lines PQ and QR. [9]

Answers

Method (M) marks are for "knowing a method and attempting to apply it."
Accuracy (A) marks can only be awarded if the relevant (M) marks have been earned.
(B) marks are independent of method marks.

(1)

$x^2 - 7x - 18 > 0$ M1

$(x - 9)(x + 2) > 0$ A1

$x > 9$ or $x < -2$ A1A1

(2) (a)   10 m; 6 m   B1B1

(b)

Use of $t = 8$ M1

$8c = \pi$ M1

$c = \pi/8$ A1

(3) (a)

$f(-2) = 0$ M1

Hence result by factor theorem   A1

(b)

$f(x) = (x + 2)(2x^2 + 2x - 1)$, by inspection   B1B1

$2x^2 + 2x - 1 = 0$

$x = \dfrac{-2 \pm \sqrt{4 + 8}}{4} = -\dfrac{1}{2} \pm \dfrac{1}{2}\sqrt{3}$ M1A2

and $x = -2$ A1

(4)

$2(1 - \cos^2 x) = 3 - 3\cos x$ M1

$2\cos^2 x - 3\cos x + 1 = 0$ M1

$(2\cos x - 1)(\cos x - 1) = 0$ A1

$\cos x = \dfrac{1}{2}$ or $\cos x = 1$ A1A1

$x = \dfrac{\pi}{3}, \dfrac{5\pi}{3}$ or $x = 0, 2\pi$ A3

(5) (a)

$6hx + 4x^2 = 300$ M1A1

$h = 50x^{-1} - \dfrac{2x}{3}$ A1

(b)

$V = 2x^2 h$ M1

$= 2x^2 \left(50x^{-1} - \dfrac{2x}{3}\right)$ M1

$= 100x - \dfrac{4x^3}{3}$ A1

(c)

$\dfrac{dV}{dx} = 100 - 4x^2$ M1A1

$100 - 4x^2 = 0$ M1

$x = \pm 5$, but $x > 0$ A1

$x = 500 - \dfrac{500}{3} = \dfrac{1000}{3}$ A1

(6) (a)

$x^2 + y^2 - 2xy$ M1

$= (x - y)^2 \geqslant 0$ A1

Hence $x^2 + y^2 \geqslant 2xy$ A1

(b)

$A - x^2 = y^2 - A$ M1

$2A = x^2 + y^2$ A1

$A = \dfrac{1}{2}(x^2 + y^2)$ A1

(c) $\dfrac{G}{x^2} = \dfrac{y^2}{G}$ M1

$G^2 = x^2 y^2$ A1

$G = xy$ (not $-xy$ as $G > 0$)   A1

(d) From **(a)**, $2G \leqslant 2A$ M1

$G \leqslant A$ A1

(e) $x = y$ B1

(7) (a) $y = 2x + 1$

$(2x + 1)^2 = 2x^2 + x$ M1A1

$2x^2 + 3x + 1 = 0$ A1

$(2x + 1)(x + 1) = 0$ A1

$x = -\dfrac{1}{2}$ or -1 A1

$y = 0$ or -1 A1

(b) Gradient of L is 2

Gradient of perpendicular is $-\dfrac{1}{2}$ M1

$y + 1 = -\dfrac{1}{2}(x + 1)$ M1A1

$2y + x + 3 = 0$ A1

(c) $(1, -2)$ lies on above line   M1

Dist. $= \sqrt{(-1 - 1)^2 + (-1 + 2)^2}$ A1

$= \sqrt{5} = 2.24$ A1

(8) (a) $3x^2 - 18x + 27 = 12$ M1

$x^2 - 6x + 5 = 0$

$(x - 1)(x - 5) = 0$ M1A1

$x = 1$ or 5

P is $(1, 12)$; Q is $(5, 12)$ A1A1

(b) R is $(0, 27)$ B1

Area of trapezium under $RQ = \dfrac{1}{2} \times (27 + 12) \times 5 = \dfrac{195}{2}$ B2

Area of rectangle under $PQ = 12 \times 4 = 48$ B1

Area under curve $RP = \displaystyle\int_0^1 3x^2 - 18x + 27\,\mathrm{d}x$ M1

$= \left[x^3 - 9x^2 + 27x\right]_0^1$ A1

$= 1 - 9 + 27 = 19$ A1

Shaded area $= \dfrac{195}{2} - (48 + 19) = 30.5$ M1A1

Mock Exam

Time: 1 hour 30 minutes

(1) Find the values of x for which

$$\frac{x-3}{x+1} = 2 - \frac{3}{x}, \, x \neq 0, -1.$$ 　　　　　　　　**[4]**

(2) The coefficient of x^3 in the expansion of $\left(3x + \frac{k}{3x}\right)^7$ is 63.

Find the values of the constant k. 　　　　　　　　**[6]**

(3) (a) Given that $y = \ln(2x) - x^2$, find $\frac{d^2 y}{dx^2}$. 　　　　**[3]**

(b) Find the exact coordinates of the stationary point on the curve
with equation $y = \ln(2x) - x^2$. 　　　　　　　　**[3]**

(c) Determine the nature of the stationary point. 　　　　**[2]**

(4) Solve, for $-\pi \leqslant x \leqslant \pi$, giving your answers in terms of π,

(a) $3\sin x + \cos 2x = 2$. 　　　　　　　　**[5]**

(b) $\sin 3x + \sin x = \sin 2x$. 　　　　　　　　**[5]**

(5) (a) Find the constant c such that
$x^2 - 4x + 9 = (x - 2)^2 + c$, for all values of x. 　　**[2]**

(b) Sketch the graph of $y = x^2 - 4x + 9$, giving the equation of its axis of
symmetry and the coordinates of its turning point. 　　**[4]**

The function f is given by
$$f : x \to x^2 - 4x + 9, \, x \, \varepsilon \, \mathbb{R}.$$

(c) Find the range of f. 　　　　　　　　**[2]**

(d) Explain, with reference to your sketch, why f has no inverse with its given
domain and suggest a domain for f for which it does have an inverse. 　　**[3]**

(6) (a) Express $5\cos x + 12\sin x$ in the form $R\sin(x + \alpha)$, where $R > 0$ and
$0° < \alpha < 90°$, giving α to 2 decimal places. 　　**[4]**

(b) Find, to 1 decimal place, the values of x, where $0° < x < 360°$, for which
$$5\cos x = 6.5 - 12\sin x.$$ 　　　　　　　　**[5]**

(c) Find the minimum value of $(5\cos x + 12\sin x)^3$. 　　**[2]**

(7) The tangent at $P(x_n, x_n^2 - 2)$, where $x_n > 0$, to the curve with equation $y = x^2 - 2$ meets the axis at the point $Q(x_{n+1}, 0)$.

(a) Show that $x_{n+1} = \dfrac{x_n^2 + 2}{2x_n}$. [6]

(b) Using the above iteration formula, with $x_1 = 2$, find x_2 and x_3 as fractions and show that $x_4 = \frac{577}{408}$. [3]

(c) The numbers x_1, x_2, x_3 and x_4 are successive approximations to the positive root of a certain equation. Find, explaining your method, this equation. [2]

(8) **(a)** Find the area of the finite region R bounded by the curve with equation $y = x(4 - 3x)$ and the straight line $y = x$. [7]

(b) Find the volume of the solid of revolution formed when the region R is rotated through one revolution about the x-axis, giving your answer in terms of π. [7]

Answers

Method (M) marks are for "knowing a method and attempting to apply it."
Accuracy (A) marks can only be awarded if the relevant (M) marks have been earned.
(B) marks are independent of method marks.

(1) $x(x - 3) = 2x(x + 1) - 3(x + 1)$ M1
$0 = x^2 + 2x - 3$ A1
$0 = (x + 3)(x - 1)$ M1
$x = -3$ or 1 A1

(2)   Coefficient of $x^3 = {}^7C_2 3^5 \left(\frac{k}{3}\right)^2$ M1A1A1
$21.27.k^2 = 63$ M1
$= \frac{1}{9}$ A1
$k = \pm\frac{1}{3}$ A1

(3) (a) $\frac{dy}{dx} = \frac{1}{x} - 2x = x^{-1} - 2x$ M1A1

$\frac{d^2y}{dx^2} = -x^{-2} - 2$ A1

(b) $\frac{dy}{dx} = x^{-1} - 2x = 0$ M1

$x = \frac{1}{\sqrt{2}}$ $(x > 0)$ A1

$y = \ln \sqrt{2} - \frac{1}{2}$ A1

(c)   When $x^2 = \frac{1}{2}$ M1

$\frac{d^2y}{dx^2} = -4$, maximum   A1

(4) (a) $3\sin x + 1 - 2\sin^2 x = 2$ M1
$2\sin^2 x - 3\sin x + 1 = 0$ A1
$(2\sin x - 1)(\sin x - 1) = 0$
$\sin x = \frac{1}{2}$ or $\sin x = 1$ A1
$x = \frac{\pi}{6}, \frac{5\pi}{6}, \frac{\pi}{2}$ A2

(b) $2\sin 2x \cos x = \sin 2x$ M1
$\sin 2x(2\cos x - 1) = 0$ M1
$\sin 2x = 0$ or $\cos x = \frac{1}{2}$ A1
$2x = -\pi, 0, \pi$ or $x = \pm\frac{\pi}{3}$
$x = \pm\frac{\pi}{2}, 0, \pm\frac{\pi}{3}$ A2

(5) (a)   Put $x = 2$, $5 = c$ M1A1

(b) $y = (x - 2)^2 + 5$
$x = 2$ is line of symmetry;   B1
$(2, 5)$ is turning point   B1

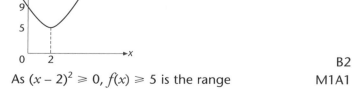

     B2

(c)   As $(x - 2)^2 \geqslant 0$, $f(x) \geqslant 5$ is the range   M1A1

(d) f is not $1 - 1$ on $x \varepsilon \mathbb{R}$ M1A1
$x \geqslant 2$ (or $x \leqslant 2$)   B1

(6) (a) $R\sin(x + \alpha) = 12\sin x + 5\cos x$

$R\cos\alpha\sin x + R\sin\alpha\cos x = 12\sin x + 5\cos x$ M1A1

$R\cos\alpha = 12; \ R\sin\alpha = 5$

$R = \sqrt{\left(5^2 + 12^2\right)} = 13;$ B1

$\alpha = \tan^{-1}\left(\frac{5}{12}\right) = 22.62°$ (2 d.p.) B1

(b) $12\sin x + 5\cos x = 6.5$

$13\sin(x + 22.62°) = 6.5$ M1

$\sin(x + 22.62°) = 0.5$

$(x + 22.62°) = 30°$ or $150°$ A1A1

$x = 7.4°$ or $127.4°$ A1A1

(c) $(-13)^3$ M1

-2197 A1

(7) (a) $\frac{dy}{dx} = 2x$ B1

$y - (x_n^2 - 2) = 2x_n(x - x_n)$ M1A1

when $y = 0, \ x = x_{n+1}$

$-(x_n^2 - 2) = 2x_n(x_{n+1} - x_n)$ M1

$x_{n+1} = x_n - (x_n^2 - 2)/2x_n$

$= (2x_n^2 - x_n^2 + 2)/2x_n$ M1

$= (x_n^2 + 2)/2x_n$ A1

(b) $x_2 = \frac{6}{4} = \frac{3}{2}$ B1

$x_3 = \frac{17}{12}$ B1

$x_4 = \frac{577}{408}$ B1

(c) Put $x_{n+1} = x_n = x$ say

$x = (x^2 + 2)/2x$ M1

$2x^2 = (x^2 + 2)$

$x^2 = 2$ A1

(8) (a) $x(4 - 3x) - x = 0$ M1

$x(3 - 3x) = 0$

$x = 0$ or 1 A1A1

$A = \int_0^1 y_{\text{TOP}} - y_{\text{BOTTOM}}dx$

$= \int_0^1 x(4 - 3x) - x \ dx$ M1

$= \int_0^1 x\{(4 - 3x) - 1\}dx$

$= \int_0^1 x(3 - 3x)dx$

$= \int_0^1 3x - 3x^2dx$ A1

$= \left[\frac{3x^2}{2} - x^3\right]_0^1$ A1

$= \frac{1}{2}$ A1

(b)   Vol. $= \pi \int_0^1 y^2 \mathrm{d}x -$ vol. of cone

$= \pi \int_0^1 x^2(4-3x)^2\mathrm{d}x - \frac{1}{3}\pi$ M1A1A1

$= \pi \int_0^1 x^2(16 - 24x + 9x^2)\,\mathrm{d}x - \frac{1}{3}\pi$ A1

$= \pi \int_0^1 (16x^2 - 24x^3 + 9x^4)\,\mathrm{d}x - \frac{1}{3}\pi$ M1

$= \pi \left[\frac{16}{3}x^3 - 6x^4 + \frac{9}{5}x^5\right]_0^1 - \frac{1}{3}\pi$ A1

$= \pi \left(\frac{16}{3} - 6 + \frac{9}{5} - \frac{1}{3}\right)$

$= \frac{4\pi}{5}$ A1

Mock Exam

Time: 1 hour 30 minutes

N.B. Only a basic scientific calculator may be used for this paper

(1) Any odd number can be written in the form $2k + 1$, where k is an integer. Prove that the difference of the squares of two odd numbers is divisible by 4. **[4]**

(2) A new car cost £10 000 and its value, £V, after t years is given by

$$V = 10\,000 \times 0.8^t.$$

Find the rate at which its value is falling when it has lost 50% of its initial value, stating the units of your answer. **[5]**

(3) (a) Expand $\sqrt{(4 + x)}$ in ascending powers of x, as far as the term in x^2. **[4]**

   (b) Write down the range of values of x for which the expansion is valid. **[2]**

(4) A curve C is given parametrically by the following equations
$$x = t^2 - 1, \, y = 2t + 2, \, t \, \varepsilon \, R.$$

The normal to this curve at the point P where $t = 2$ meets the curve again at the point Q. Find the value of t at Q.

[9]

(5) A circle C, with centre O, has equation
$$x^2 + y^2 - 10y + 16 = 0.$$

   (a) Find the coordinates of O. **[2]**

   (b) Show that the radius of C is 3. **[1]**

The point P (2.4, 3.2) lies on C. The line L is the tangent to C at P.

   (c) Find the equation of L. **[5]**

   (d) Verify that the point Q (–3, –4) lies on L. **[1]**

   (e) Find the size of angle $O\hat{Q}P$. **[3]**

(6) The vectors **p** and **q** are given by

$$\mathbf{p} = \lambda\,\mathbf{i} + (2\lambda - 1)\mathbf{j} - \mathbf{k}$$
$$\mathbf{q} = (1 - \lambda)\mathbf{i} + 3\lambda\mathbf{j} + (4\lambda - 1)\mathbf{k}$$

where λ is a scalar.

(a) Find the values of λ for which **p** and **q** are perpendicular. **[6]**

When $\lambda = 2$, **p** and **q** are the position vectors of the points P and Q respectively, relative to a fixed origin O.

(b) Find $\overrightarrow{PQ}$. **[2]**

(c) Find, to the nearest degree, the size of angle QPO. **[5]**

(7) (a) By using the substitution $u^2 = x - 1$, or otherwise, find

$$\int \frac{x+1}{\sqrt{(x-1)}}\,\mathrm{d}x.$$

[7]

(b) Find $\int x\cos 3x\,\mathrm{d}x$. **[4]**

(c) Hence evaluate $\displaystyle\int_{0}^{\frac{\pi}{6}} x\cos 3x\,\mathrm{d}x$ **[2]**

(8) (a) Write $\dfrac{1}{x(3-x)}$ in partial fractions. **[3]**

(b) Find $\displaystyle\int \frac{\mathrm{d}x}{x(3 - x)}$. **[3]**

(c) Solve the differential equation

$$3x\,\frac{\mathrm{d}y}{\mathrm{d}x} = y(3 - y)$$

given that when $x = 2$, $y = 2$, expressing y as a function of x. **[7]**

Answers

Method (M) marks are for "knowing a method and attempting to apply it."
Accuracy (A) marks can only be awarded if the relevant (M) marks have been earned.
(B) marks are independent of method marks.

(1)

$(2k + 1)^2 - (2m + 1)^2$ M1
$= (2k + 1 + 2m + 1)(2k + 1 - 2m - 1)$
$= (2k + 2m + 2)(2k - 2m)$ A1
$= 4(k + m + 1)(k - m)$ M1
hence divisible by 4   A1

(2)

$\frac{dV}{dt} = 10\ 000 \times \ln 0.8 \times 0.8^t$ M1A1
$\quad = \ln 0.8 V$ A1
When $V = 5000$,
$\frac{dV}{dt} = 5000 \times \ln 0.8 = -1115.72$ M1
i.e. falling at £1115.72 per year   A1

(3) (a)

$\sqrt{4\left(1 + \frac{x}{4}\right)}$ M1
$= 2\left(1 + \frac{x}{4}\right)^{1/2}$ M1
$= 2\left(1 + \left(\frac{1}{2}\right)\left(\frac{x}{4}\right) + \frac{\left(\frac{1}{2}\right)\left(-\frac{1}{2}\right)\left(\frac{x}{4}\right)^2}{2!}\right)$ A1
$= 2 + \frac{x}{4} - \frac{x^2}{64}$ A1

(b)

$\left|\frac{x}{4}\right| < 1$ M1
$|x| < 4$ A1

(4)

$\frac{dy}{dt} = 2; \frac{dx}{dt} = 2t$
$\frac{dy}{dx} = \frac{2}{2t} = \frac{1}{t}$ M1A1
Gradient of normal is $-t$
$y - 6 = -2(x - 3)$ M1A1
$y = -2x + 12$
$2t + 2 = -2(t^2 - 1) + 12$ M1A1
$0 = 2t^2 + 2t - 12$
$0 = t^2 + t - 6$ M1
$0 = (t - 2)(t + 3)$
$t = 2$ or -3 A1
Answer is -3 A1

(5) (a)

$x^2 + y^2 - 10y + 16 = 0$
$x^2 + y^2 - 10y + 25 = 25 - 16$
$x^2 + (y - 5)^2 = 9 = 3^2$ M1
Centre is (0, 5)   A1

(b) Radius is 3   A1

(c)

$2x + 2y\frac{dy}{dx} - 10\frac{dy}{dx} = 0$ M1
$\frac{dy}{dx} = \frac{x}{(5-y)}$ A1
$y - 3.2 = \left(\frac{2.4}{1.8}\right)(x - 2.4)$ M1A1
$3y = 4x$ A1

(d) $3 \times (-4) = 4 \times (-3)$ B1

(e) $PQ = 9$ B1
       $\tan OQP = \frac{3}{9}$ M1
       $OQP = 18.4°$ A1

(6) (a) $\lambda(1 - \lambda) + (2\lambda - 1)3\lambda - (4\lambda - 1) = 0$ M1A1
         $5\lambda^2 - 6\lambda + 1 = 0$ A1
         $(5\lambda - 1)(\lambda - 1) = 0$ M1
         $\lambda = \frac{1}{5}$ or 1   A1A1

(b) $\lambda = 2$: $\mathbf{p} = 2\mathbf{i} + 3\mathbf{j} - \mathbf{k}$
       $\mathbf{q} = -\mathbf{i} + 6\mathbf{j} + 7\mathbf{k}$ M1
       $\mathbf{PQ} = -3\mathbf{i} + 3\mathbf{j} + 8\mathbf{k}$ A1

(c) $\mathbf{PQ.PO}$
       $= (-3\mathbf{i} + 3\mathbf{j} + 8\mathbf{k})(-2\mathbf{i} - 3\mathbf{j} + \mathbf{k})$ M1
       $= 5$ A1
       Also $\mathbf{PQ.PO} = \sqrt{82}\,\sqrt{14}\,\cos QPO$ M1
       $\cos QPO = \frac{5}{\sqrt{1148}}$ A1
       $QPO = 82°$ (nearest degree)   A1

(7) (a) $u^2 = x - 1 \rightarrow 2u\frac{\mathrm{d}u}{\mathrm{d}x} = 1$ M1A1

         $\int \frac{(u^2 + 2)}{u}\, 2u\mathrm{d}u$ M1A1

         $= 2\int (u^2 + 2)\mathrm{d}u$

         $= 2(\frac{u^3}{3} + 2u) + c$ A1

         $= 2\frac{u}{3}(u^2 + 6) + c$ M1

         $= \frac{2\sqrt{(x-1)}}{3}(x + 5) + c$ A1

(b) $\int x\cos 3x\mathrm{d}x$

       $= \frac{1}{3}x\sin 3x - \int \frac{1}{3}\sin 3x\mathrm{d}x$ M1A1A1

       $= \frac{1}{3}x\sin 3x + \frac{1}{9}\cos 3x + c$ A1

(c) $\int\limits_{0}^{\pi/6} x\cos 3x\mathrm{d}x$

       $= \left[\frac{1}{3}x\sin 3x + \frac{1}{9}\cos 3x\right]_{0}^{\pi/6}$

       $= \frac{1}{18}\pi - \frac{1}{9}$ M1A1

(8) (a)   Let $\frac{1}{x(3-x)} = \frac{A}{x} + \frac{B}{(3-x)}$

         $1 = A(3 - x) + Bx$ M1
         $x = 0$: $A = \frac{1}{3}$ A1
         $x = 3$: $B = \frac{1}{3}$ A1

(b) $\frac{1}{3} \int \frac{1}{x} + \frac{1}{(3-x)} \, dx$

$= \frac{1}{3}(\ln x - \ln(3-x)) + c$ 　　　　　M1A1A1

$= \frac{1}{3} \ln \frac{Ax}{(3-x)}$

(c) $3 \int \frac{dy}{y(3-y)} = \int \frac{dx}{x}$ 　　　　　M1

$\ln \frac{Ay}{(3-y)} = \ln x$ 　　　　　A1

$\frac{Ay}{(3-y)} = x$ 　　　　　A1

$x = 2, y = 2$

$2A = 2 \rightarrow A = 1$ 　　　　　M1

$y = x(3-y)$ 　　　　　A1

$y + xy = 3x$ 　　　　　M1

$y = \frac{3x}{(1+x)}$ 　　　　　A1